Springer Series in Materials Science

Volume 347

The Springer Series in Materials Science covers the complete spectrum of materials research and technology, including fundamental principles, physical properties, materials theory and design. Recognizing the increasing importance of materials science in future device technologies, the book titles in this series reflect the state-of-the-art in understanding and controlling the structure and properties of all important classes of materials.

Daisuke Shindo · Zentaro Akase

Electron Holography for Electromagnetic Materials

Springer

Daisuke Shindo
Center for Emergent Matter Science
RIKEN
Wako, Japan

Zentaro Akase
Institute of Multidisciplinary research for
Advanced Materials
Tohoku University
Sendai, Japan

Data Science Center
Nara Institute of Science and Technology
Ikoma, Japan

ISSN 0933-033X ISSN 2196-2812 (electronic)
Springer Series in Materials Science
ISBN 978-981-97-3536-5 ISBN 978-981-97-3537-2 (eBook)
https://doi.org/10.1007/978-981-97-3537-2

This Springer imprint is published by the registered company Springer Nature Singapore Pte Ltd.
The registered company address is: 152 Beach Road, #21-01/04 Gateway East, Singapore 189721, Singapore

Preface

To fully understand the properties of electromagnetic materials, it is necessary to characterize the electromagnetic fields in and around such materials. By utilizing the electron interference effect, electron holography provides valuable information about electric and magnetic fields on the nanometer scale. Thus, to understand and utilize electron holography techniques effectively, basic knowledge is needed about electromagnetic fields and interference effects derived from quantum mechanics and the theory of relativity. This book presents the fundamental theories of and the latest developments in electron holography.

Chapter 1 explains Maxwell's equations and electromagnetic potentials on the basis of the special theory of relativity. This theoretical basis is useful to understanding the principles and application of electron holography discussed in Chaps. 2 and 3, respectively. In addition to relativistic kinematics, Einstein's field equations based on general relativity are presented. These formulations and quantum mechanics are important to the discussions in Chaps. 4 and 5 of the complicated behaviors and wave-particle dualism of the electron, respectively.

Chapter 2 outlines the basic principles and formulations of electron holography. In addition to an explanation of the concept of holography, the detailed data analysis procedures used in electron holography are presented. An explanation is also given of how magnetic information can be separated from electric information.

Chapter 3 describes the application of electron holography to various materials with classification in terms of various shapes such as particles, lines, and plates. In addition, electron holography analysis of bulk electromagnetic materials and devices is presented. Furthermore, a description is given of the electromagnetic field and collective motion of electrons obtained by in situ observation of the effects of applying various voltages and magnetic fields.

Chapter 4 describes the effects of dynamical diffraction and strong deflection on electron holography analysis. Phase contrast variations due to the dynamical diffraction effect are explained on the basis of computer simulation results. The characteristic features of the strong deflection effect on reconstructed phase images are explained in terms of the wave-particle dualism of the electron, which is considered in the following chapter.

Chapter 5 theoretically considers the interference effect and point-charge behavior of the electron, i.e., its wave-particle behavior, on the basis of experimental results obtained using electron holography observation of the electromagnetic field and collective motion of electrons. It also presents the spinning linear wave (SLW) model of the electron, which takes into account the relevant quantum mechanics and the general theory of relativity, and describes the diffraction and interference effects of the electron simulated using the SLW model. Finally, Larmor precession of the spin magnetic moment of the electron is interpreted in terms of the SLW model.

To visualize various dynamic variations of electromagnetic field and Larmor precession of spin angular moment with the SLW model, videos and animations are presented.

Wako, Japan Daisuke Shindo
Sendai, Japan Zentaro Akase

About This Book

Electron holography provides valuable information about the properties of electromagnetic materials. This book presents fundamental theories and latest developments in electron holography. In addition to Maxwell's equations and electromagnetic potentials on the basis of the special theory of relativity, for understanding the wave behavior of the electron, Einstein's field equations based on general relativity are explained. Application of electron holography to various materials with classification in terms of various shapes such as particles, lines, and plates including bulks and devices is presented. Not only electromagnetic field but also collective motion and spin polarization of electrons are visualized. SLW (Spinning Linear Wave) model originated from holography data and the theory of relativity is successfully applied to interpret quantum phenomena, such as diffraction, interference, and spin of the electron. Included are videos and animations visualizing various dynamic variations of electromagnetic field and Larmor precession of spin angular moment simulated with the SLW model.

Contents

Chapter 1
Theoretical Basis

Abstract Maxwell's equations and electromagnetic potentials are explained on the basis of the special theory of relativity. This theoretical basis is useful to understanding the principles and application of electron holography discussed in Chaps. 2 and 3, respectively. In addition to relativistic kinematics, Einstein's field equations based on general relativity are presented. These formulations and quantum mechanics are important to the discussions in Chaps. 4 and 5 of the complicated behaviors and wave-particle dualism of the electron, respectively.

This chapter explains Maxwell's equations and electromagnetic potentials on the basis of the special theory of relativity to facilitate understanding of the principles and application of electron holography described in the following chapters. Einstein's field equations based on general relativity are presented as well as the relativistic kinematics of the electron. These formulations and quantum mechanics are important to the discussions of the complicated behaviors and wave-particle dualism of the electron in Chaps. 4 and 5, respectively.

1.1 Maxwell's Equations and Special Theory of Relativity

1.1.1 Maxwell's Equations and Electromagnetic Potentials

We begin by sorting out Maxwell's equations and their relationships with electromagnetic potentials. Maxwell's equations for electric charge density ρ and electric current density $\overrightarrow{j}$ are given by

$$\mathrm{div}\,\vec{B} = 0, \tag{1.1}$$

$$\frac{\partial \vec{B}}{\partial t} + \mathrm{rot}\,\vec{E} = 0, \tag{1.2}$$

D. Shindo and Z. Akase, *Electron Holography for Electromagnetic Materials*, Springer Series in Materials Science 347, https://doi.org/10.1007/978-981-97-3537-2_1

$$\text{div } \vec{D} = \rho, \tag{1.3}$$

$$\text{rot } \vec{H} - \frac{\partial \vec{D}}{\partial t} = \vec{j}. \tag{1.4}$$

In the above equations, $\vec{D}$ and $\vec{E}$ are the electric flux density and electric field, respectively, and their relationship is given by

$$\vec{D} = \varepsilon_0 \vec{E}. \tag{1.5}$$

On the other hand, $\vec{B}$ and $\vec{H}$ are the magnetic flux density and magnetic field, respectively, and their relationship is given by

$$\vec{B} = \mu_0 \vec{H}. \tag{1.6}$$

In (1.5) and (1.6), ε_0 and μ_0 are respectively the dielectric constant and magnetic permeability for vacuum, i.e.,

$$\varepsilon_0 \mu_0 = 1/c^2. \tag{1.7}$$

From (1.1) and (1.2), $\vec{B}$ and $\vec{E}$ are given with scalar and vector potentials ϕ and $\vec{A}$, i.e.,

$$\vec{B} = \text{rot } \vec{A}, \tag{1.8}$$

$$\vec{E} = -\frac{\partial \vec{A}}{\partial t} - \text{grad}\varphi. \tag{1.9}$$

If we insert (1.8) and (1.9) into (1.3) and (1.4), respectively, we obtain

$$\Box \varphi = \rho / \varepsilon_0. \tag{1.10}$$

$$\Box \vec{A} = \mu_0 \vec{j}, \tag{1.11}$$

where $\Box$ is the d'Alenbertian given with

$$\Box = \left(\frac{\partial}{c\partial t}\right)^2 - \left(\frac{\partial}{\partial x}\right)^2 - \left(\frac{\partial}{\partial y}\right)^2 - \left(\frac{\partial}{\partial z}\right)^2. \tag{1.12}$$

Equations (1.10) and (1.11) are equivalent to (1.3) and (1.4) under the Lorenz condition [1];

$$\mathrm{div}\vec{A} + \frac{\partial \rho}{c^2 \partial t} = 0. \tag{1.13}$$

This is justified, since from (1.10) and (1.11),

$$\Box\,(\mathrm{div}\,\vec{A} + \frac{\partial \varphi}{c^2 \partial t}) = \mu_0(\mathrm{div}\vec{j} + \frac{\partial \rho}{\partial t}), \tag{1.14}$$

and we notice that the right side of equation (1.14) is zero from the charge conservation law.

1.1.2 Maxwell's Equations Formulated Using Special Theory of Relativity

The electric field around a charged particle at rest is evaluated using (1.3). On the other hand, when the particle or observer moves, the magnetic field is formed and thus (1.4) should be taken into account.

In the special theory of relativity, the physical laws hold in their simplest forms for any coordinate system subjected to a uniform translational motion relative to another coordinate system called as Lorentz covariance. Under this condition and with the same light velocity for both systems, Maxwell's equations are reformulated as follows.

In order to specify a point in the space–time of physics, a four-coordinate representation $x^\mu(\mu = 0, 1, 2, 3)$ is used:

$$x^0 = ct, \quad x^1 = x, \quad x^2 = y, \quad x^3 = z. \tag{1.15}$$

A quantity with a suffix written in the upper position is called a "contravariant element." Let us take the coordinates of the point be $x + dx^\mu$ near the point close to x^μ. The displacement written with the four coordinates are considered to be a vector called a "contravariant vector." The squared magnitude of line element ds:

$$(ds)^2 = (dx^0)^2 - (dx^1)^2 - (dx^2)^2 - (dx^3)^2 \tag{1.16}$$

is independent of the orientation and the local coordinate system and is called a "scalar." Equation (1.16) is in accordance with Dirac's notations [2]. For generalization, this equation is rewritten as

$$(ds)^2 = \sum_{\mu=0}^{3} \sum_{\nu=0}^{3} \eta_{\mu\nu} dx^\mu\, dx^\nu, \tag{1.17}$$

where $\eta_{\mu\nu}$ is called the Minkowskian metric tensor and is given as

$$\eta_{\mu\nu} = \eta_{\nu\mu} = \left\{ \begin{array}{c} 1(\mu = \nu = 0) \\ 0\ (\mu \neq \nu) \\ -1\ (\mu = \nu = 1, 2, 3) \end{array} \right\}. \tag{1.18}$$

The tensor has the suffix written in the lower position and is called a "covariant tensor." Following Einstein [3], the summation symbol is omitted and thus (1.17) is simply written as

$$(ds)^2 = \eta_{\mu\nu} dx^{\mu} dx^{\nu}. \tag{1.19}$$

In the following, when we have a formula term with the same suffix in both the upper and lower positions, we consider it to be a summation taken over four values, i.e., from 0 to 3.

The contravariant vector A^{μ} can be related to the scalar and vector potentials as

$$A^0 = \varphi/c,\ \ A^1 = A_x,\ \ A^2 = A_y,\ \ A^3 = A_z. \tag{1.20}$$

By using $\eta_{\mu\nu}$, contravariant vector A^{μ} is transformed into covariant vector A_{μ}:

$$A_{\mu} \equiv \eta_{\mu\nu} A^{\nu}. \tag{1.21}$$

Here, we note

$$A_0 = A^0 = \varphi/c,\ \ A_k = -A^k\ \ (k = 1,\ 2,\ 3). \tag{1.22}$$

Similarly, a four-current j^{μ} consisting of charge density (ρ) and current (j_x, j_y, j_z) is described as

$$j^0 = c\rho,\ \ j^1 = j_x,\ \ j^2 = j_y,\ \ j^3 = j_z. \tag{1.23}$$

Using these notations, two of Maxwell's equations (1.10) and (1.11) are given as

$$\Box A^{\lambda} = j^{\lambda} \tag{1.24}$$

where

$$\Box \equiv \eta_{\mu\nu} \frac{\partial}{\partial x^{\mu}} \frac{\partial}{\partial x^{\nu}}. \tag{1.25}$$

The Lorenz condition [1] can be written as

$$\partial_{\mu} A^{\mu} = 0, \tag{1.26}$$

where

$$\partial_\mu = \frac{\partial}{\partial x^\mu}. \tag{1.27}$$

To represent electric and magnetic fields, we introduce an antisymmetric covariant tensor of the 2nd rank:

$$f_{\mu\nu} = -f_{\nu\mu} = \frac{\partial A_\nu}{\partial x^\mu} - \frac{\partial A_\mu}{\partial x^\nu} = \partial_\mu A_\nu - \partial_\nu A_\mu. \tag{1.28}$$

This gives us the correspondence between $f_{\mu\nu}$ and $\left(\vec{E}, \vec{B}\right)$:

$$\begin{pmatrix} f_{00}\, f_{01} & f_{02}\, f_{03} \\ f_{10}\, f_{11} & f_{12}\, f_{13} \\ f_{20}\, f_{21} & f_{22}\, f_{23} \\ f_{30}\, f_{31} & f_{32}\, f_{33} \end{pmatrix} = \begin{pmatrix} 0 & E_x/c & E_y/c & E_z/c \\ -E_x/c & 0 & -B_z & B_y \\ -E_y/c & B_z & 0 & -B_x \\ -E_z/c & -B_y & B_x & o \end{pmatrix}. \tag{1.29}$$

Thus, the first two of Maxwell's equations (1.1 and 1.2) are given by

$$\partial_\lambda f_{\mu\nu} + \partial_\mu f_{\nu\lambda} + \partial_\nu f_{\lambda\mu} = 0. \tag{1.30}$$

On the other hand, the second two of Maxwell's equations (1.3 and 1.4) are given by

$$\partial_\nu f^{\lambda\nu} = \mu_0 j^\lambda. \tag{1.31}$$

These equations reformulated in accordance with the special theory of relativity hold in the same form for any coordinate system subjected to a uniform translational motion relative to another coordinate system (Lorentz covariance).

The coordinates in inertia system *S*', which has velocity *v* in relation to inertia system *S* in direction x^1 (=*x*), are given with the Lorentz transformation as

$$x'^\mu = a^\mu{}_\nu x^\nu, \tag{1.32}$$

where $a^\mu{}_\nu$ is the coefficient corresponding to the Lorentz transformation; i.e.,

$$\begin{pmatrix} a^0{}_0\, a^0{}_1 & a^0{}_2\, a^0{}_3 \\ a^1{}_0\, a^1{}_1 & a^1{}_2\, a^1{}_3 \\ a^2{}_0\, a^2{}_1 & a^2{}_2\, a^2{}_3 \\ a^3{}_0\, a^3{}_1 & a^3{}_2\, a^3{}_3 \end{pmatrix} = \begin{pmatrix} 1/\sqrt{1-\beta^2} & -\beta/\sqrt{1-\beta^2} & 0 & 0 \\ -\beta/\sqrt{1-\beta^2} & 1/\sqrt{1-\beta^2} & 0 & 0 \\ 0 & 0 & 1 & 0 \\ 0 & 0 & 0 & 1 \end{pmatrix} \tag{1.33}$$

$$\left(\beta = \frac{v}{c}\right).$$

With these reformulations of Maxwell's equations formulated in accordance with the special theory of relativity, Maxwell's energy–momentum tensor $T^{\mu\nu}$ is given as

$$T^{\mu\nu} = \frac{1}{\mu_0}(\eta_{\lambda\sigma} f^{\lambda\nu} f^{\sigma\mu} - \frac{1}{4}\eta^{\nu\mu} f^{\alpha\beta} f_{\alpha\beta}) = T^{\nu\mu}. \tag{1.34}$$

This formulation is used for the discussion in Sect. 1.3.

1.2 Relativistic Kinematics

As shown in the previous section, Maxwell equations are consistent with the special theory of relativity. Equations (1.30) and (1.31), which are reformulated in accordance with the special theory of relativity, shows the form of Lorentz invariance.

In contrast, Newton's law satisfies the Galilean transformation and is not consistent with the special theory of relativity. When a particle is considered to be a point mass, Newton's law can be modified to satisfy the special theory of relativity, as shown below.

It should be noted that the field around a particle is precisely treated using Einstein's field equations on the basis of the general theory of relativity, as discussed in the next section. Newton's law can be obtained as an approximate form of Einstein's field equations. Furthermore, whether a particle is considered to be a point mass is closely related to wave-particle dualism, which will be discussed in detail in Chap. 5. This is important for mathematical treatments of electron holography data.

While time t in the Lorentz transformation depends on the coordinates in each inertia system that satisfy (1.19), t is independent of the coordinates in the Galilean transformation. In Newtonian mechanics, the kinematics of a point mass is represented by the coordinates of the point mass x^k ($k = 1, 2, 3$) as a function of time t. From the relativistic point of view, this treatment is not adequate since time and space are not treated equivalently. Here we introduce "proper time τ," given by

$$\begin{aligned} d\tau &= \frac{1}{c}(\eta_{\mu\nu} dx^\mu dx^\nu)^{\frac{1}{2}} \\ &= dt\sqrt{1 - (\frac{\vec{v}}{c})^2}, \end{aligned} \tag{1.35}$$

where $\vec{v}$ is the velocity of the point mass. From (1.35), $d\tau$ satisfies the Lorentz invariance. At proper time τ, let us observe this point mass in coordinate system S', which moves at the same velocity as the point mass. Since the observed point mass is still in coordinate system S', from (1.35) with $\vec{v} = 0$,

$$d\tau = dt'. \tag{1.36}$$

Thus, the proper time corresponds to a clock assigned to the point mass.

With proper time τ, the four-velocity of the point mass is given by a contravariant vector, i.e.,

$$u^{\mu} = \frac{dx^{\mu}(\tau)}{d\tau}. \tag{1.37}$$

Let's consider the point mass fixed to the coordinate origin of system S'. The coordinates of this point observed from systems S' and S are $x^{\mu'}(\tau) = (x^{0\prime}, \vec{x}' = 0)$, $x^{\mu}(\tau)$.

The transformations between $x^{\mu'}(\tau)$ and $x^{\mu}(\tau)$ are given by

$$x^{\mu\prime} = a^{\mu}{}_{\nu} x^{\nu} \tag{1.38}$$

$$x^{\nu} = b^{\nu}{}_{\mu} x^{\mu'} = b^{\nu}{}_{0} x^{0'}, \tag{1.39}$$

where we have

$$b^{\nu}{}_{\mu} = a_{\mu}{}^{\nu}. \tag{1.40}$$

When we divide (1.39) by τ,

$$\frac{dx^{\nu}(\tau)}{d\tau} = u^{\nu} = b^{\nu}{}_{0} c \frac{dt'(\tau)}{d\tau} = c\, b^{\nu}{}_{0}. \tag{1.41}$$

On the other hand, when we divide (1.39) by t,

$$v^{k} = \frac{dx^{k}}{dt} = b^{k}{}_{0} \frac{dx^{o'}}{dt} \quad (k = 1,\ 2,\ 3) \tag{1.42}$$

and

$$c = \frac{dx^{0}}{dt} = b^{0}{}_{0} \frac{dx^{o'}}{dt}. \tag{1.43}$$

Furthermore,

$$\eta_{00} = \eta_{\mu\nu} b^{\mu}{}_{0} b^{\nu}{}_{0} = [c^{2} - (\vec{v})^{2}]/(\frac{dx^{0'}}{dt})^{2}, \tag{1.44}$$

and thus, from (1.42),

$$b^{k}{}_{0} = \frac{v^{k}}{(\frac{dx^{0'}}{dt})} = \frac{v^{k}}{c\sqrt{1-\beta^{2}}} \tag{1.45}$$

and from (1.43)

$$b^0{}_0 = \frac{c}{(\frac{dx^{0'}}{dt})} = \frac{1}{\sqrt{1-\beta^2}}. \tag{1.46}$$

From (1.41), we have

$$u^k = \frac{v^k}{\sqrt{1-\beta^2}} \tag{1.47}$$

and

$$u^0 = \frac{c}{\sqrt{1-\beta^2}}. \tag{1.48}$$

The four-momentum is given by

$$p^\mu = mu^\mu, \tag{1.49}$$

where p^μ consists of the momentum of the point mass,

$$p^k = \frac{mv^k}{\sqrt{1-\beta^2}}, \tag{1.50}$$

and the energy of the point mass,

$$E = cp^0 = \frac{mc^2}{\sqrt{1-\beta^2}}. \tag{1.51}$$

If $\vec{v}$ is 0, $\vec{p}$ is 0 from (1.50). On the other hand, energy E is mc^2 from (1.51) even if $\vec{v}$ is 0. The mc^2 is called the rest energy for mass m and is very important to discussions of the law of energy conservation. The rest energy of the electron is used for introducing de Broglie's matter wave, as discussed in Sect. 1.4.1.

In the derivation of the momentum and energy for the point mass in (1.50) and (1.51), respectively, a particle was assumed to be the point mass, and Newton's law was modified to satisfy the special theory of relativity. As noted above, the field around a particle can be precisely treated using Einstein's field equations on the basis of the general theory of relativity, as discussed in the next section. Also as noted above, whether a particle is considered to be a point mass will be discussed in Chap. 5. This will be done on the basis of Einstein's field equations, taking into account the spinning motion.

1.3 Outline of Einstein's Field Equations Based on General Theory of Relativity

The above discussion of Maxwell's equation and the relativistic mechanics of the electron was based of the special theory of relativity. To discuss the gravitational field around an electron in detail, it is necessary to consider Einstein's field equations in the framework of the general theory of relativity with a four-dimensional Riemannian space. The squared magnitude of line element *ds* given by Minkowskian metric tensor $\eta_{\mu\nu}$ defined in (1.18) for the Minkowskian space can be generalized to metric tensor $g_{\mu\nu}$ for the Riemannian space:

$$(ds)^2 = g_{\mu\nu} dx^{\mu} dx^{\nu}, \tag{1.52}$$

The special theory of relativity is considered to be a special case of the general theory of relativity in which $g_{\mu\nu}$ have constant values given by (1.18). By considering the case in which $g_{\mu\nu}$ differ from constant values by small magnitudes compared with unity, Newton's law of universal gravitation can be derived [3].

According to Newton's law, the gravitational potential φ is given by the following Poisson equation

$$\Delta\varphi = 4\pi G\rho, \tag{1.53}$$

where G and ρ are the universal gravitational constant and the density of matter. In the relativistic theory of gravitation, $\Delta\varphi$ is replaced by the contracted curvature tensor $R^{\alpha}_{\mu\nu}(= R^{\alpha}_{\mu\nu\alpha})$, which is obtained by contraction of Riemann-Christoffel curvature tensor$R^{\alpha}{}_{\mu\nu\beta}$; i.e.,

$$R^{\alpha}_{\mu\nu\beta} = \frac{\partial}{\partial x^{\beta}}\Gamma^{\alpha}_{\mu\nu} - \frac{\partial}{\partial x^{\nu}}\Gamma^{\alpha}_{\mu\beta} - \Gamma^{\tau}_{\mu\nu}\Gamma^{\alpha}_{\tau\beta} + \Gamma^{\tau}_{\mu\beta}\Gamma^{\alpha}_{\tau\nu}, \tag{1.54}$$

where

$$\Gamma^{\alpha}_{\mu\nu} = -\frac{1}{2}g^{\alpha\sigma}\left(\frac{\partial g_{\mu\sigma}}{\partial x^{\nu}} + \frac{\partial g_{\nu\sigma}}{\partial x^{\mu}} - \frac{\partial g_{\mu\nu}}{\partial x^{\sigma}}\right). \tag{1.55}$$

On the right side of (1.53), we need a tensor in place of ρ. According to the special theory of relativity, mass corresponds to energy; we accordingly set energy–momentum tensor $T_{\mu\nu}$ except that belonging to the pure gravitational field on the right side. Thus, Einstein's field equations are given by [4, 5]

$$R_{\mu\nu} - \frac{1}{2}Rg_{\mu\nu} = -\kappa T_{\mu\nu} \quad (\mu,\ \nu = 0,\ 1,\ 2,\ 3), \tag{1.56}$$

where the second term on the left is added due to the formal reason in such a way that its divergence disappears identically in the sense of absolute differential calculus.

Note that R is the curvature scalar ($R = g^{\mu\nu}R_{\mu\nu}$) and κ is Einstein's constant of gravitation given as $8\pi G/c^4$. Since these equations are nonlinear partial differential equations in terms of $g_{\mu\nu}$, they are generally difficult to solve.

Following the special theory of relativity in 1905 [6] and the general theory of relativity in 1915 [3, 7], Einstein discussed the stability of an electron in 1919 [8]. He modified (1.56) to take into account the charge by including Maxwell's energy–momentum tensor (1.34) for the right side of (1.56). By assuming that the stability of the particle is solely due to gravity and that Maxwell's energy–momentum tensor is proportional to some quantity consisting of $g_{\mu\nu}$ and their derivatives, the following relation was obtained:

$$f_{\mu\nu}j^{\nu} + \frac{1}{4\kappa}\frac{\partial R}{\partial x^{\mu}} = 0, \tag{1.57}$$

where $f_{\mu\nu}$ is an antisymmetric covariant tensor corresponding to the electric and magnetic field components given by (1.28) and j^{ν} is a contravariant four current given by (1.23). In (1.57), the first term corresponds to the Coulomb repulsive force and the second term is the gravitational compressive force. Thus, the equation shows that the gravitational compressive force and the Coulomb repulsive force are balanced.

In Chap. 5, we discuss Einstein's field equations further on the basis of the approximate solution obtained by Infeld and Schild [9]. In that discussion, the spinning linear wave (SLW) model [10] is used to represent wave-particle dualism. The Larmor precession of the electron spin magnetic moment is also discussed using this model.

1.4 Disturbance-Free Observation Based on Quantum Mechanics

1.4.1 De Broglie's Matter Wave and Wave Function

To understand the principles of electron holography in Chap. 2, we utilized the wave function extensively based on the quantum mechanics. Before the introduction of the wave function, de Broglie postulated the wave nature of the electron in 1923 [11]. First de Broglie assumed the relation

$$mc^2 = h\nu, \tag{1.58}$$

where m is the rest mass of the electron. The left side of this equation corresponds to the rest energy of the electron, which is obtained using (1.51) with $\beta^2\ (= (\frac{\vec{v}}{c})^2) = 0$. Then, he considered an electron with a stationary harmonic oscillation represented by $\sin(2\pi\nu t\prime)$, where ν is the frequency and t' is time [11, 12]. By applying the Lorentz transformation to the harmonic oscillation at a particle position for an observer in an inertia system moving with speed v in the z-direction, de Broglie derived [11]

$$\sin(2\pi\nu t') = \sin[\frac{2\pi\nu}{\sqrt{1-\frac{v^2}{c^2}}}\left(t - \frac{vz}{c^2}\right)]$$
$$= \sin(\omega t - kz)$$
$$(\omega = \frac{2\pi\nu}{\sqrt{1-\frac{v^2}{c^2}}},\ k = 2\pi/\lambda), \quad (1.59)$$

where λ (= h / p; h: Planck's constant) is the de Broglie wavelength. The last line of (1.59) represents the matter wave propagating in the z-direction.

Based on the de Broglie's matter wave, a plane wave of incident electrons propagating in the z-direction is represented in quantum mechanics as a wave function in complex notation as given by

$$\psi_{PW}(\vec{z}, t) = \exp[i(\vec{k}\cdot\vec{z} - \omega t)]. \quad (1.60)$$

1.4.2 Disturbance-Free Observation

On the basis of quantum mechanics, we discuss the electron holographic visualization of collective motions of electrons (Chap. 3) utilizing the wave function discussed above [13].

In general, the orbits or trajectories of electrons and other particles have been macroscopically investigated using inelastic scattering processes. Inelastic experiments processes cause the decoherence [14]. In other words, the measurement apparatus disturbs the system to be observed. On the other hand, in electron holography, the measurement apparatus comprises elastic electrons, and the object to be observed is the electric and magnetic fields of the system, which consists of a charged specimen and secondary electrons.

As explained in Chap. 3, a stationary state can be assumed in which the charged specimen and the motions of secondary electrons around it are static since the velocities of the incident electrons are about 100 times higher than those of secondary electrons (Sect. 3.1.2). Therefore, the interaction of the incident electrons with the system consisting of the charged specimen and the secondary electrons is given by the time-independent Schrödinger equation with the Hamiltonian H as

$$(-\frac{\hbar^2}{2m}\nabla^2 + H_s + H_{\text{int}})\Psi = E\Psi, \quad (1.61)$$

where the three terms in parentheses correspond respectively to the kinetic energy of the incident electrons, the energy of the system consisting of the charged specimen plus secondary electrons (H_s), and the interaction energy (H_{int}) between the incident electrons and the system. The total wave function for the incident electron and the system consisting of the charged specimen and secondary electrons can be expanded in a series:

$$\Psi(\vec{r}, \vec{R}_j, \vec{R}_k) = \sum_n \psi_n(\vec{r}) D_n(\vec{R}_j, \vec{R}_k), \tag{1.62}$$

In this equation, $\psi_n(\vec{r})$ is the wave function of the incident electron, and $\psi_0(\vec{r})$ and $\psi_{n\neq 0}(\vec{r})$ are the elastic scattering and the inelastic scattering, respectively. D_n is the wave function of the system representing the nth excited state with energy ε_n, and $\overrightarrow{R}_{j,k}$ represents the positions of the constituent charges (electrons (j) and nuclei (k)). D_n is given as

$$H_s D_n = \varepsilon_n D_n. \tag{1.63}$$

Here, we note that the energy loss of the inelastically scattered electrons with wave function $\psi_n(\vec{r})$ is given as

$$\Delta E = E - E_n = \varepsilon_n. \tag{1.64}$$

From (1.62) and (1.61), a set of the following equations for $\psi_n(\vec{r})$:

$$\left[-\frac{\hbar}{2m}\nabla^2 - E_n + H_{nn}\right]\psi_n = -\sum_{m\neq n} H_{nn}\psi_m, \quad n = 0,\ 1,\ \ldots, \tag{1.65}$$

where H_{nm} is the matrix element given by

$$\begin{aligned} H_{nm}(\vec{r}) &= \int D_n^*(\vec{R}_j,\ \vec{R}_k) H_{\mathrm{int}}(\vec{r}, \vec{R}_j, \vec{R}_k) D_m(\vec{R}_j,\ \vec{R}_k) d^3R_j d^3R_k \\ &= \langle D_n | H_{\mathrm{int}} | D_m \rangle. \end{aligned} \tag{1.66}$$

In the above equation, the off-diagonal elements H_{nm} corresponds to the probability of an inelastic transition from $(D_m,\ \psi_m)$ to $(D_n,\ \psi_n)$ as discussed in detail by Yoshioka [15].

As expressed by (1.66), the inelastically scattered electrons $\psi_n(\vec{r})(\mathrm{n} \neq 0)$ affect the observed intensity in conventional electron microscopy disturbing the system. On the other hand, holograms are obtained from only $\psi_0(\vec{r})$ through the interference effect in electron holography. Here we note that inelastically scattered wave $\psi_n(\vec{r})(n \neq 0)$ cannot contribute to the interference fringes. Thus, by simplifying (1.65), $\psi_0(\vec{r})$ is given by

$$(-\frac{\hbar}{2m}\nabla^2 - E_0 + H_{00})\psi_0 = 0, \tag{1.67}$$

where H_{00} is potential energy $-e\varphi$ (φ: electric potential). The solutions to (1.67) are two components of $\psi_0(\vec{r})$, i.e., the object wave and the reference wave.

In the field of quantum measurement, as disturbance-free measurements, the quantum non-demolition measurements [16–18] and negative-result measurements [19] for photons and spins have been studied. Electron holographic visualization of collective motions of secondary electrons can be carried out without disturbing their motions. In addition, it should be noted that the density of secondary electrons observed by electron holography is much larger than that of incident electrons.

References

1. L.V. Lorenz, Phil. Mag. Ser. **3**(34), 287 (1867). https://doi.org/10.1080/14786446708639882
2. P.A.M. Dirac, *Geneal Theory of Relativity*. Wiley (1975)
3. A. Einstein, Ann. der Phys. Ser. **4**(49), 769 (1916)
4. C. Møller, *The Theory of Relativity* (Clarendon Press, Oxford, Oxford, 1952)
5. W. Pauli, *Theory of Relativity* (Pergamon Press, London, 1958)
6. A. Einstein, *Zur Elecktrodynamik bewegter Körper, Ann.der Phys.* **17**, 891 (1905)
7. A. Einstein, *Zur allgemeinen Relativitätstheorie*, S.B. Press. Akad. Wiss, 777 (1915)
8. A. Einstein, *Spielen Gravitationsfelder im Aufbau der materierllen Elementareilen ein wesentliche Rolle?*,Berlin Berichte, 349 (1919)
9. L. Infeld, A. Schild, Rev. Mod. Phys. **21**, 408 (1949)
10. D. Shindo, T. Tomita, *Material Characterization using Electron Holography* (Wiley-VCH, Weinheim, 2023)
11. L. de Broglie, Ondes et Quanta. Comptes Rendus **177**, 507 (1923)
12. G. Lochak, *De Broglie's initial conception of de Broglie waves*, in The Wave-Particle Dualism, ed.by S. Diner, D. Fargue, G. Lochak, and F. Selleri, D. Reidel, 1 (1984)
13. D. Shindo, J. Jung Kim, K. Hyun Kim, W. Xia, N. Ohno, Y. Fujii, N. Terada, S. Ohno, J. Phys. Soc. Jpn. **78**, 104802 (2009)
14. M. Schlosshauer, Rev. Mod. Phys. **76**, 1267 (2005)
15. H. Yoshioka, J. Phys. Soc. Japan **12**, 618 (1957)
16. V.B. Braginsky, Y.I. Vorontsov, F.Y. Khalili, *Zh. Eksp. Teor. Fiz* **73**, 1340 (1977)
17. V.B. Braginsky, Y.I. Vorontsov, K.S. Thorne, Science **209**, 547 (1980)
18. P. Grangier, J.A. Levenson, J.-P. Poizat, Nature **396**, 537 (1998)
19. M. Renninger, Z. Phys. **158**, 417 (1960)

Chapter 2
Principles and Practice of Electron Holography

Abstract Basic principles and formulations of electron holography are described. In addition to an explanation of the concept of holography, the detailed data analysis procedures used in electron holography are presented. An explanation is also given of how magnetic information can be separated from electric information.

2.1 What is Holography?

The word "holography" originates from the Greek words "holos" and "graphia," which mean "whole" and "recording," respectively. Thus, "holography" means "complete recording." The concept of holography was introduced by Gabor in 1948 [1]. To improve the resolution of electron microscopes, Gabor aimed to correct the spherical aberration of the objective lens by using an interference microscope and a method for reconstructing wavefronts. However, since highly coherent electron beams were not available at that time, this idea was not used to improve electron microscopes.

The invention of the laser enabled applications of holography to be developed in the field of light [2]. Figure 2.1 illustrates the process for recording an optical hologram by using a laser beam. While conventional photographs present two-dimensional projected amplitude images, the holographic reconstruction process produce three-dimensional images using the following process. As indicated in Fig. 2.1, the laser beam is split into two beams with a beam splitter. One part of the beam illuminates the target to produce an object wave. The object wave then interferes with the reference wave to form a hologram on a photographic plate. The image on the plate contains phase information corresponding to the distance between the object and the plate. By illuminating the plate with a reconstruction beam, a diffracted beam is generated from the photographic plate, as shown in Fig. 2.2. Since

Supplementary Information The online version contains supplementary material available at https://doi.org/10.1007/978-981-97-3537-2_2.

D. Shindo and Z. Akase, *Electron Holography for Electromagnetic Materials*, Springer Series in Materials Science 347, https://doi.org/10.1007/978-981-97-3537-2_2

the diffracted beam contains the amplitude and phase information, the same perspective view as the original object wave can be seen when the diffracted wave is observed from the right with both eyes.

In electron holography, the phase shift of the incident electron wave due to the electromagnetic field can be detected by utilizing the wave feature of the electron, and the electric and magnetic fields can be visualized as explained in detail below. Figure 2.3a illustrates a typical bar magnet. It attracts magnetic materials due to the magnetic field around it. But we cannot see such a field with the naked eye, even with conventional microscopy. As explained later, with electron holography, such

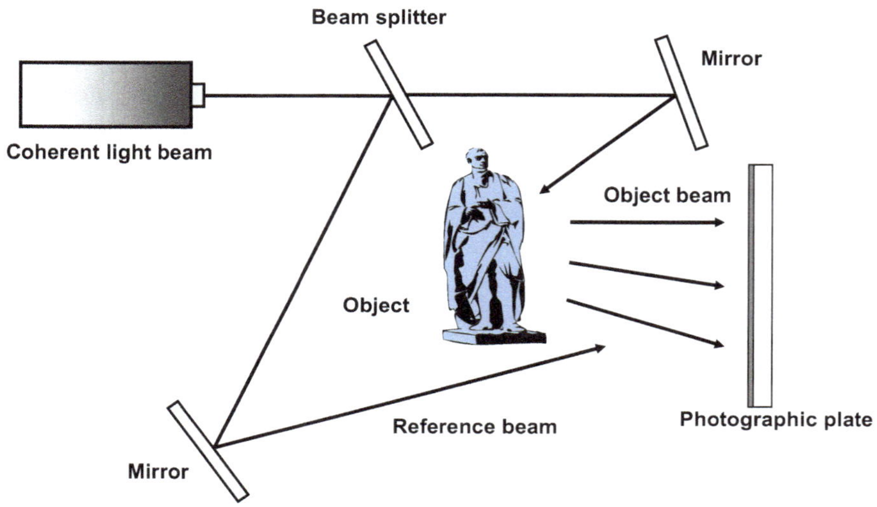

Fig. 2.1 Illustration showing formation of optical hologram

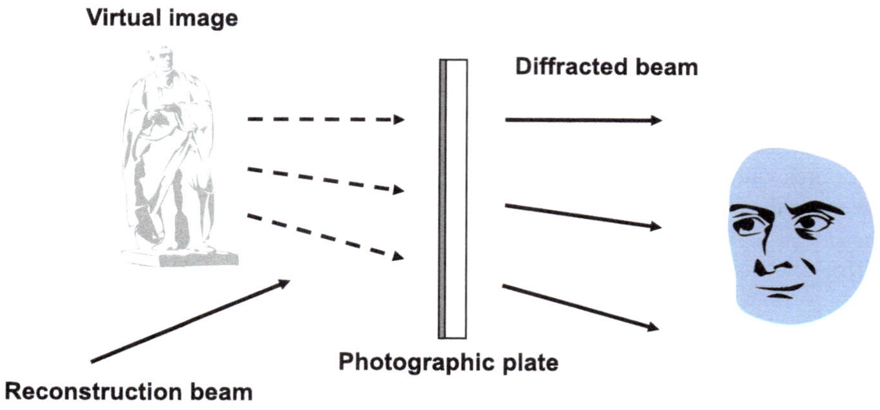

Fig. 2.2 Illustration showing reconstruction of virtual image in optical holography

magnetic fields can be visualized as shown in Fig. 2.3b, where the magnetic field around the magnet is simulated.

Figure 2.4a shows an electron microscope image of a piece of Co–Zr–O magnetic material [3]. To record the magnetic field information, a hologram is obtained utilizing the interference effect of the electron wave, as shown in Fig. 2.4b. Utilizing the Fourier transformation enables the magnetic field not only around but also inside the specimen to be visualized, as shown in Fig. 2.4c.

Gabor originally proposed using an interference microscope and carried out in-line holography, which is a simple way of recording a hologram. The basic geometrical configuration of in-line holography is illustrated in Fig. 2.5a where the transmitted plane wave acts as a reference wave. Both the object wave (green region) and the reference wave propagate parallel to the optical axis. This approach has limitations, however. First, the spatial coherence length is insufficient to form a hologram covering a wide region. Second, inelastic scattering forms the background. Further, the specimen should be small compared with the region of the reference wave.

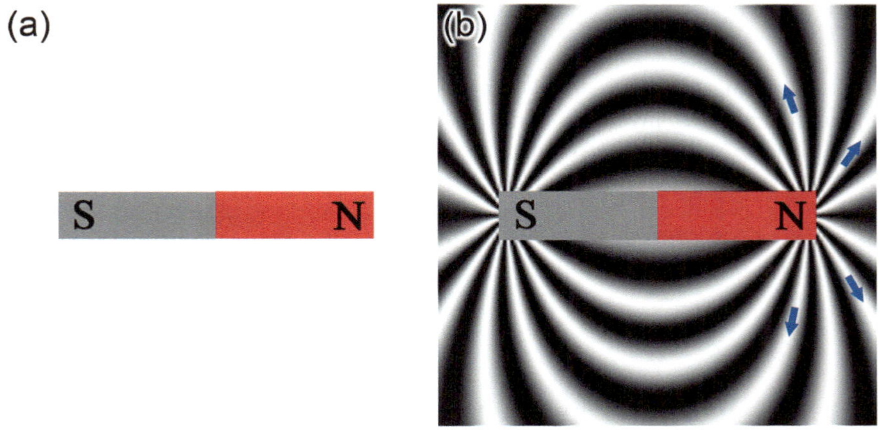

Fig. 2.3 **a** Bar magnet. **b** Simulated magnetic field around magnet. Arrows indicate direction of magnetic flux

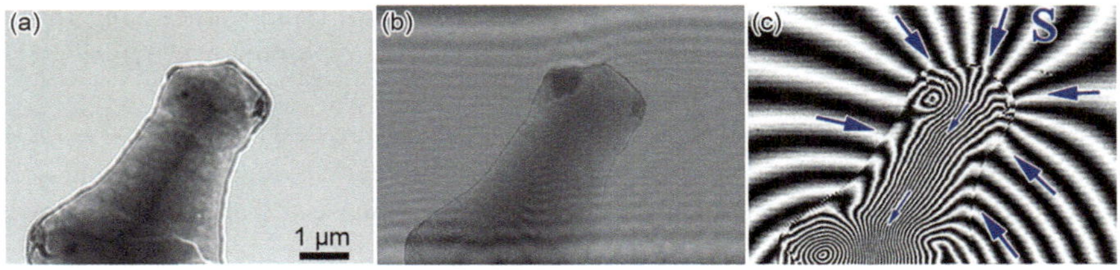

Fig. 2.4 **a** Electron microscope image showing piece of Co–Zr–O magnetic specimen. **b** Hologram obtained through interference effect of incident electrons. **c** Reconstructed phase image showing detailed magnetic fields both inside and outside material

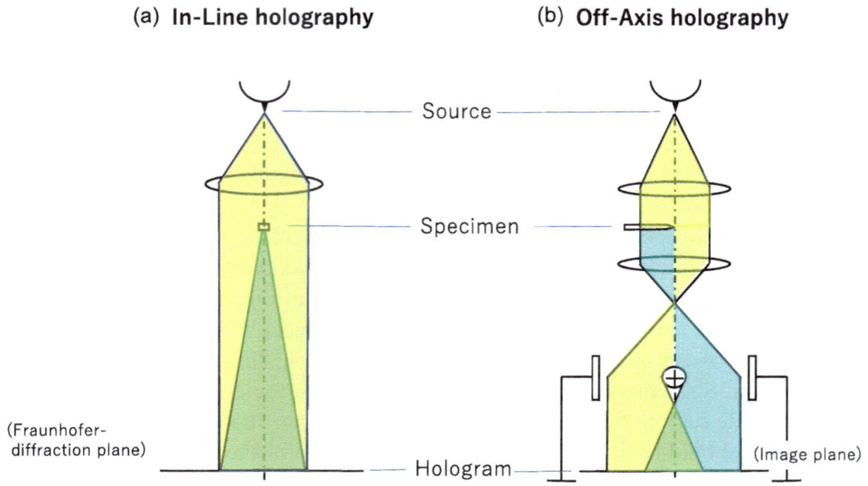

Fig. 2.5 Two types of electron holography: **a** in-line and **b** off-axis

To overcome the geometrical limitations of in-line holography, off-axis electron holography was invented by introducing a biprism was introduced as shown in Fig. 2.5b, and extensively developed by Möllendstedt and his colleagues at Tübingen University [4]. In off-axis electron holography, the object wave (blue region) passing through the specimen and the reference wave are definitely separated. This technique has been widely utilized these days. Since the following explanations are made with the latter method, "off-axis electron holography" is referred to as "electron holography" for short.

2.2 Fourier Transformation and Sign Conventions

2.2.1 Fourier Transformation

To understand the formulation of electron scattering and image processing in the following sections, it is important to understand the correspondence between the electron scattering process and the Fourier transformation.

The plane wave of an incident electron is generally expressed by $\exp[i(\vec{k}\cdot\vec{r}-\omega t)]$, where k and ω are the wave number and angular frequency given by $2\pi/\lambda$ and $2\pi\nu$ with λ: wavelength and ν: frequency, respectively. The phase and amplitude changes of the plane wave for a specimen in the stationary condition are independent of time. Under this condition, the incident plane wave is simply given by $\exp(i\vec{k}\cdot\vec{r})$. When the effect of the specimen on the plane wave is expressed by $q(x, y)$, the scattering amplitude at observation point (s, t) (see Fig. 2.6a) is given as

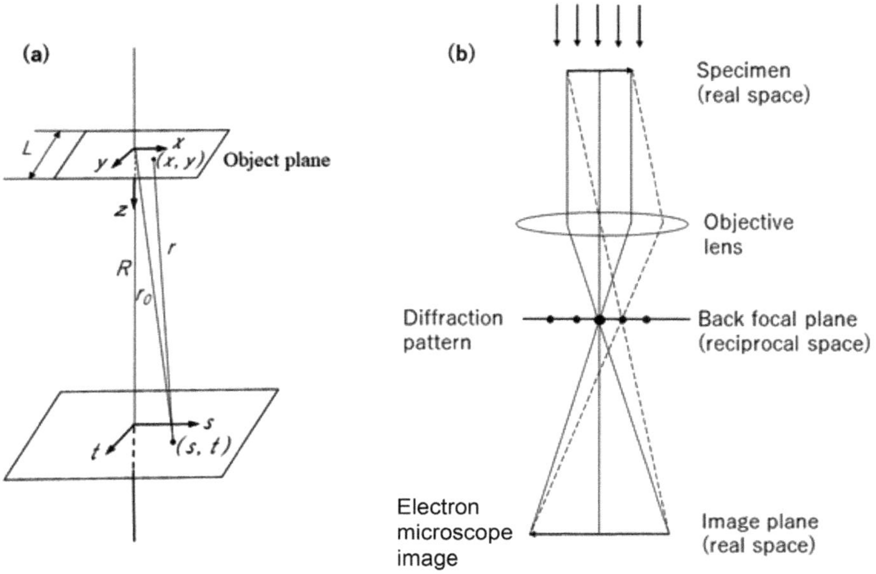

Fig. 2.6 **a** Schematic showing electron scattering in a specimen. **b** Diagram of imaging process in transmission electron microscope

$$\psi(s, t) = C \iint q(x, y) \frac{\exp(i\,kr)}{r}\, dx\, dy, \tag{2.1}$$

where C is a constant, and r is given as

$$r = \sqrt{R^2 + (x - s)^2 + (y - t)^2}. \tag{2.2}$$

The first term in the square root is generally larger than the sum of the second and third terms, and thus the binomial theorem is applied to r; i.e.,

$$\begin{aligned} r &= R\sqrt{1 + \frac{(x - s)^2 + (y - t)^2}{R^2}} \\ &= R + \frac{(x - s)^2 + (y - t)^2}{2R} - \frac{[(x - s)^2 + (y - t)^2]^2}{8R^3} \cdots \\ &= R + \frac{s^2 + t^2}{2R} - \frac{sx + ty}{R} + \frac{x^2 + y^2}{2R} - \frac{[(x - s)^2 + (y - t)^2]^2}{8R^3} \cdots . \end{aligned} \tag{2.3}$$

Inclusion of up to the 3rd term is referred to as "Fraunhofer approximation" while that up to 4th term is "Fresnel approximation." The approximation is specified by comparing the value of $[k\,(x^2 + y^2)/2R]$ with $\pi/2$. If the specimen area is an L-square, $[k\,(x^2 + y^2)/2R]$ has a value of $\pi/2$ for $R = L^2/\lambda$, and thus

$$R > \frac{L^2}{\lambda} \quad \text{(Fraunhofer approximation)}, \tag{2.4}$$

$$R < \frac{L^2}{\lambda} \quad \text{(Fresnel approximation)}. \tag{2.5}$$

In the Fraunhofer approximation, (2.3) is simply given by

$$r \simeq r_0 - sx/r_0 - ty/r_0. \tag{2.6}$$

Thus, (2.1) can be rewritten as

$$\psi(u, v) = C' \iint q(x, y)\exp[-2\pi i(ux + vy)]dx\, dy, \tag{2.7}$$

where C', u, and v are given, respectively, as

$$C' = C \exp(ik\, r_0)/r_0, \quad u = s/\lambda r_0, \quad v = t/\lambda r_0. \tag{2.8}$$

The right side of (2.7) is a form of the Fourier transformation and thus the scattering amplitude $\psi(u, v)$ is obtained with the Fourier transformation [5, 6]. The conditions for Fraunhofer diffraction are satisfied for electron diffraction with an electron lens system in an electron microscope as illustrated in Fig. 2.6b.

To illustrate the properties of Fourier transformation, the relationship between a two-dimensional object $q(x, y)$ and the amplitude of its Fourier transformation $|\psi(u, v)| = \left|\mathscr{F}\left[q(x, y)\right]\right|$ up is described.

For the circular holes in Fig. 2.7a, b,

$$\begin{aligned} q_1(x, y) = 1,\ & \sqrt{x^2 + y^2} \leq r \\ 0,\ & \sqrt{x^2 + y^2} > r, \end{aligned} \tag{2.9}$$

the amplitude of their Fourier transformation is given by

$$|\psi_1(u, v)| = \pi r^2 \left|J_1\left(2\pi r\sqrt{u^2 + v^2}\right)\right| / (2\pi r\sqrt{u^2 + v^2}) \tag{2.10}$$

(J_1: Bessel function of the first kind of order 1), as shown in Fig. 2.7c, d.

For the rectangular holes in Fig. 2.8a, b,

$$\begin{aligned} q_2(x, y) = 1,\ & |x| \leq \frac{a}{2} \ \mathit{and}\ |x| \leq \frac{b}{2} \\ 0,\ & |x| > \frac{a}{2} \ \mathit{or}\ |x| > \frac{b}{2}, \end{aligned} \tag{2.11}$$

the amplitude of their Fourier transformation is given by

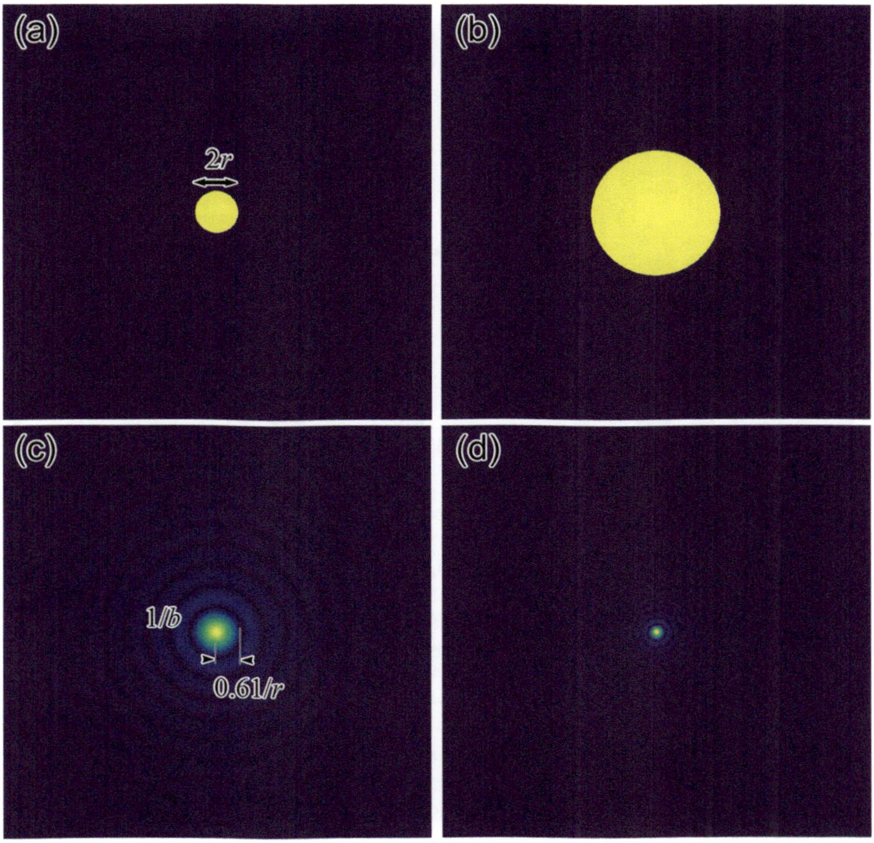

Fig. 2.7 Simulated diffraction amplitude for holes with radius r: **a** r is small; **b** r is large; **c** diffraction amplitude of (**a**); (**d**) diffraction amplitude of (**b**)

$$|\psi_2(u, v)| = \left| ab \frac{\sin \pi au}{\pi au} \cdot \frac{\sin \pi bv}{\pi bv} \right|, \tag{2.12}$$

as shown in Fig. 2.8c, d. The relationships of (2.9) and (2.10) for the case of a $\gg$ b are useful for understanding the formation of electron diffraction patterns on the basis of the spinning linear wave model described in Sect. 5.1.4.

2.2.2 *Sign Conventions*

In the above discussion, we used $\exp[i(\vec{k} \cdot \vec{r} - \omega t)]$ as the wave function for an incident electron. This function satisfies the Schrödinger wave equation; i.e.,

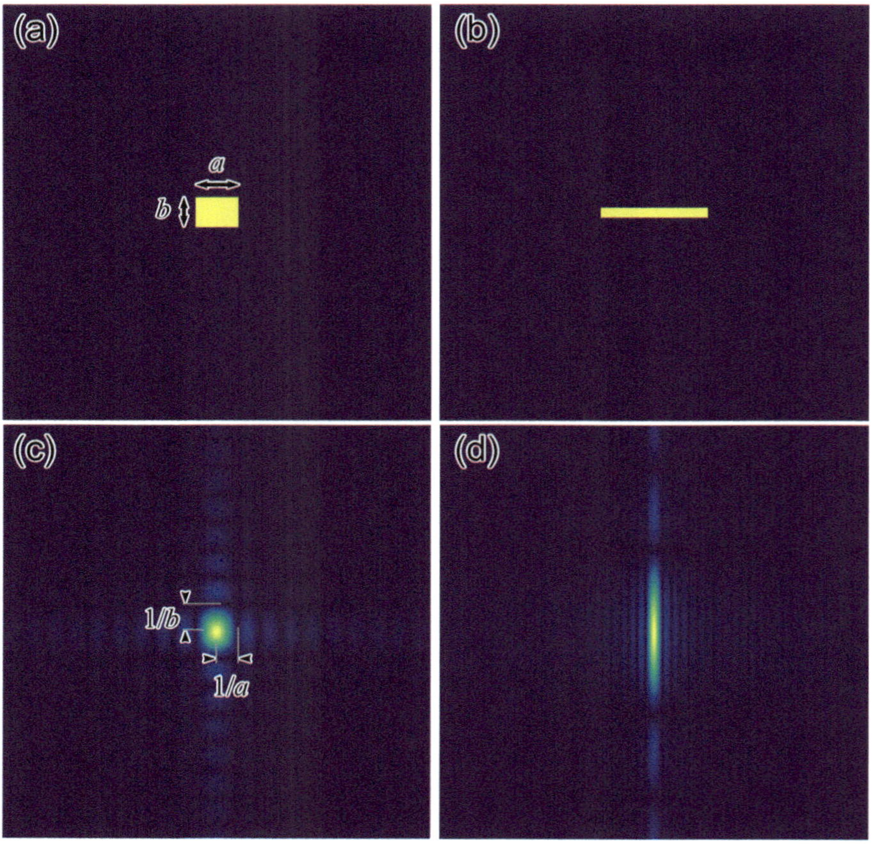

Fig. 2.8 Simulated diffraction amplitude of rectangular holes with horizontal length a and vertical length b: **(a)** a is slightly larger than b; **b** a is much larger than b; **c** diffraction amplitude of (**a**); **d** diffraction amplitude of (**b**)

$$i\hbar\frac{\partial\psi}{\partial t} = -\frac{\hbar^2}{2m}\frac{\partial^2\psi}{\partial\vec{r}^2}. \tag{2.13}$$

On the basis of this form of the wave function, the form of the transmission function, that of the Fourier transformation, and the structure factor should be represented as indicated in the first column of Table 2.1. The choices of these signs [7] are called the "quantum mechanical convention." The choices of the opposite signs in the second column of Table 2.1 are the "alternative convention." These signs do not satisfy the Schrödinger wave equation, but they have been used for a long time, especially in the field of optics, as exemplified by the publication of the International Tables for X-Ray Crystallography [8]. They do not contradict the quantum mechanical formulation if the formulation does not include the time. In computer simulation of electron holography, the choice of the appropriate sign conventions should be

Table 2.1 Sign conventions

	Quantum–mechanical convention	Alternative convention
Incident wave	$\exp\left[i\left(\vec{k}\cdot\vec{r}-\omega t\right)\right]$	$\exp\left[-i\left(\vec{k}\cdot\vec{r}-\omega t\right)\right]$
Transmission function	$\exp\left[i\sigma\varphi(\vec{r})\Delta z\right]$	$\exp\left[-i\sigma\varphi(\vec{r})\Delta z\right]$
Fourier transformation (Real space to reciprocal space)	$\int\psi(\vec{r})\exp(-2\pi i\vec{u}\cdot\vec{r})d\vec{r}$	$\int\psi(\vec{r})\exp(2\pi i\vec{u}\cdot\vec{r})d\vec{r}$
Inverse Fourier transformation (Reciprocal space to real space)	$\int\Psi(\vec{u})\exp(2\pi i\vec{u}\cdot\vec{r})d\vec{u}$	$\int\Psi(\vec{u})\exp(-2\pi i\vec{u}\cdot\vec{r})d\vec{u}$
Structure factor	$\sum_j f_j\exp(-2\pi i\vec{u}\cdot\vec{r}_j)$	$\sum_j f_j\exp(2\pi i\vec{u}\cdot\vec{r}_j)$

confirmed to ensure accurate interpretation of the phase shift due to the electric and magnetic fields.

2.3 Basis of Electron Holography

Figure 2.9a illustrates the basic geometric configuration of electron holography. Figure 2.9b illustrates the two-step imaging process [9–17]. First, by interfering the object wave and the reference wave, a hologram is formed through the biprism in the electron microscope. Next, the phase shift of the incident electron is extracted from the hologram by using a computer for executing a Fourier transformation.

With a field emission gun as listed in Table 2.2, a highly coherent electron beam can be obtained and collimated to illuminate a thin specimen through a condenser lens system. Comparing with thermionic emission electron guns, field emission guns have high brightness, a small source size, and small energy spread.

When the voltage of the biprism is zero, as shown in Fig. 2.10a, a conventional electron microscope image of the absorption-diffraction contrast [18] is obtained. When the voltage is increased, interference fringes form, as shown in (b). When the voltage is further increased, the interference fringes cover the specimen region and are used to record the electric and magnetic information, as shown in (c).

The object is assumed to be illuminated by a plane wave with a unit amplitude, i.e., $\exp[i(kz-\omega t)]$. The change in the scattering amplitude of the incident plane wave due to the object is given as

$$q(x, y) = a(x, y)\exp\{i\phi(x, y)\}, \tag{2.14}$$

where $a(x, y)$ and $\phi(x, y)$ correspond to the amplitude change and phase shift, respectively [18]. Amplitude function $a(x, y)$ represents absorption and/or diffraction in the specimen. The phase shift is expressed as

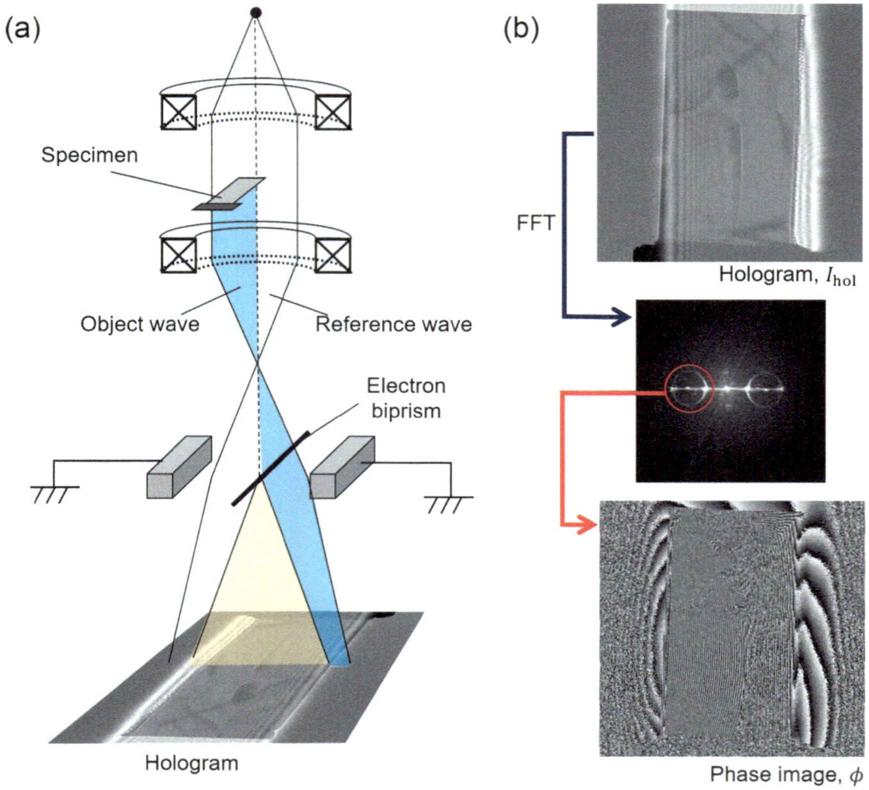

Fig. 2.9 **a** Illustration showing geometric configuration of electron holography; **b** two-step imaging process

$$\phi(\mathrm{x}, \mathrm{y}) = \sigma \int \varphi(x, y, \mathrm{z}) dz - \frac{\mathrm{e}}{\hbar} \int \vec{A}(x, y, \mathrm{z}) d\vec{z}, \tag{2.15}$$

where $\varphi(x, y, z)$ and $\vec{A}(x, y, z)$ represent the electric (scalar) and magnetic vector potentials, respectively, and σ is the interaction constant, which is determined from velocity v of the incident electrons:

$$\sigma = \frac{e}{\hbar v} = \frac{2\pi}{\lambda V\left(1 + \sqrt{1 - (v/c)^2}\right)}, \tag{2.16}$$

where V is the accelerating voltage of the electron microscope, respectively [19, 20].

In conventional electron microscopy as illustrated in Fig. 2.10a, the electron intensity is given by

Table 2.2 Characteristics of various 200-kV electron guns

			Field emission		
	Thermionic emission		Thermal field emission gun (FEG)		Cold FEG
Characteristic	W	LaB_6	$ZrO/W^{(100)}$	$W^{(100)}$	$W^{(310)}$
Brightness at 200 kV (A $(cm^2 \cdot sr)^{-1}$)	~5 × 10^5	~5 × 10^6	~5 × 10^8	~5 × 10^8	~1 × 10^9
Source size	50 μm	10 μm	0.2~1 μm	~100 nm	~100 nm
Energy spread (eV)	2.3	1.5	0.6~0.8	0.6~0.8	0.3~0.5
Current (μA)	~100	~20	~100	20~30	10~30
Stability for short time	1%	1%	1%	5%	5%
Stability for long time	1%/h	3%/h	1%/h	6%/h	5%/15 min
Current efficiency	30%	10%	10%	10%	1%
Maintenance	No need	No need	Need some times for setup	Build up several times for a new tip	Flashing necessary every few hours

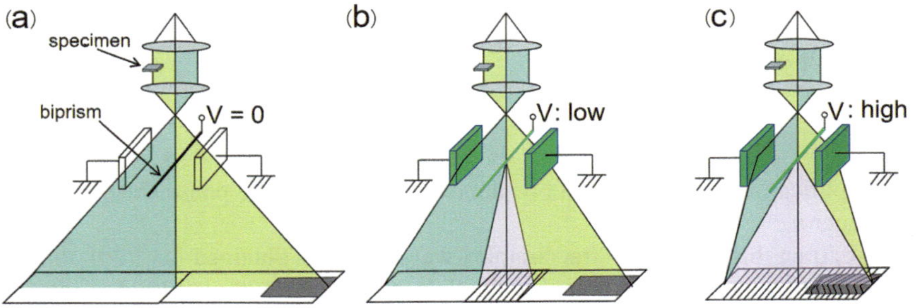

Fig. 2.10 Formation of interference fringes with increasing biprism voltage V: **a** zero, **b** low, **c** high

$$I(x, y) = |q(x, y)|^2 = a^2(x, y), \tag{2.17}$$

which indicates that only the absorption-diffraction contrast is determined. This means that the phase information, consisting of the electric and magnetic vector potentials, cannot be obtained.

In electron holography (illustrated in Fig. 2.9), on the other hand, a hologram $I_{hol}(x, y)$ is formed with the interference between the object wave and the reference wave (see also Fig. 2.4b):

$$I_{hol}(x,y) = \left| a(x,y)\exp\left(-\pi i\frac{\alpha_h}{\lambda}x + i\phi(x,y)\right) + \exp\left(\pi i\frac{\alpha_h}{\lambda}x\right)\right|^2$$
$$= 1 + a^2(x,y) + 2a(x,y)\cos\left(\frac{2\pi\alpha_h x}{\lambda} - \phi(x,y)\right), \qquad (2.18)$$

where α_h is a superimposed angle between the object wave and the reference wave, and the last term corresponds to the interference fringes. By making Fourier transformation operation ($\mathscr{F}$) on the hologram, we obtain

$$\mathscr{F}[I_{hol}(x,y)] = \delta(x,y) + \mathscr{F}[a^2(x,y)]$$
$$+ \mathscr{F}[a(x,y)\exp(i\phi(x,y))] * \delta(u + \alpha_h/\lambda, v)$$
$$+ \mathscr{F}[a(x,y)\exp(-i\phi(x,y))] * \delta(u - \alpha_h/\lambda, v), \qquad (2.19)$$

where $*$ indicates convolution operator. The third and fourth terms, which are called the side bands, on the right side of (2.19) containing information about the phase shift and amplitude change. By selecting one of the side bands such as the third term in (2.19), shifting it by α_h/λ, and performing inverse Fourier transformation ($\mathscr{F}^1$) on this term, we obtain

$$\mathscr{F}^{-1}[\mathscr{F}[a(x,y)\exp(i\phi(x,y))] * \delta(u, v)] = a(x,y)\exp(i\phi(x,y)). \qquad (2.20)$$

Thus, the phase shift and the amplitude change can be obtained. The intensity of the reconstructed phase image $I_{\mathrm{ph}}(x, y)$ is usually given in terms of the cosine function of $\phi(x, y)$:

$$I_{ph}(x,y) = \cos(\phi(x,y)). \qquad (2.21)$$

Amplification of the phase can be done by multiplying the phase shift $\phi(x, y)$ by an integer n.

To extract the phase from the complex wave function obtained in (2.20), the real and imaginary parts of the complex number at each pixel are substituted into the "arctangent 2 (atan2)" function, which is also called "atan2(y, x)" or "arctan2(y, x)." The atan2 function calculates the angle (in radians) between the positive x-axis and the ray from the origin to the point (x, y) on the Euclidean plane except point (0, 0). When the real and imaginary parts of the wave function are substituted into the atan2 function, the resulting phase is limited from $-\pi$ to π, thereby creating a map of the phase shift wrapped by 2π. Unlike the map of the phase shift created by the above cosine function, this map of the phase shift appears as a gradient in the direction of the slope of the phase, so that the direction of the gradient of the potential (e.g., see Figs. 3.6b–h, 3.10b and 3.69a) and the direction of the gradient of the vector potential (e.g., see Figs. 3.34c and 3.75a–d) can be easily read from the phase map.

The process of continuously connecting the discontinuous phases wrapped by 2π is called "the unwrapping process." The unwrapping process using the phase unwrapping algorithm enables visualization of the distribution of the overall phase shift (e.g., see Fig. 3.16a, b). This unwrapped phase shift map is used for calculation in the "procedure for separating the magnetic field-derived phase shift from the electric field-derived phase shift," which will be discussed in Sect. 2.3.3.

The distribution of equipotential lines/magnetic flux lines projected in the z-direction can be visualized by drawing contour lines on the map of the electric/magnetic field-derived phase shifts or by wrapping it with a cosine function or modulo function, which shows the discontinuous phase distribution wrapped by 2π or integer multiples of 2π (e.g., see Figs. 3.14a and 3.15a).

The direction and magnitude of the phase tilt can be calculated from the partial derivative of the phase distribution. By expanding the partial derivatives, a color map can be obtained using the HSV color space, i.e., the combination of hue value, saturation value, and brightness value, respectively (e.g., see Fig. 3.17f–h). In addition, the color map can be superimposed on the phase reconstructed image wrapped by the cosine function as a means to express the electric flux/magnetic flux distribution in an intuitive and easy-to-understand manner (e.g., see Fig. 3.16c).

2.4 Principles of Electron Holography

2.4.1 Phase Shift Due to Scalar Potential

If a specimen is a non-magnetic material, the phase shift results from the scalar potential. The phase shift is obtained from (2.15),

$$\phi(x, y) = \sigma \int \varphi(x, y, z)dz. \tag{2.22}$$

As shown in Fig. 2.11a, for a specimen with thickness t, the phase shifts of the incident electron wave at positions C and D below the specimen are compared. If the specimen is homogeneous, from (2.22), the phase difference of the phase shifts at positions C and D is given as

$$\begin{aligned} \phi(x_C, y_C) - \phi(x_D, y_D) &= \sigma \int_0^t \varphi(x_C, y_C, z) - \varphi(x_D, y_D, z)dz \\ &= 0. \end{aligned} \tag{2.23}$$

For such a condition, the phase shift remains the same, and thus information about the inner potential cannot be obtained.

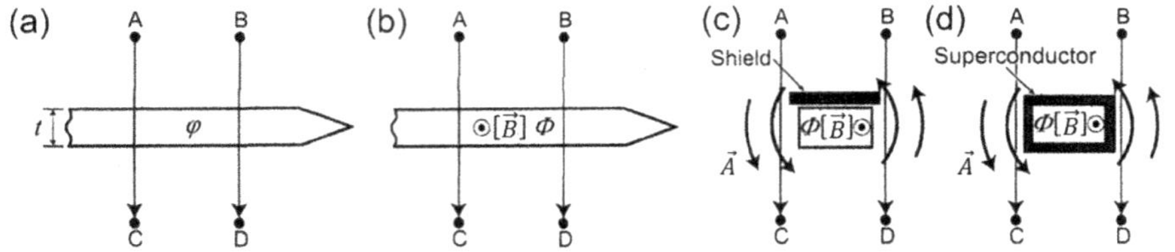

Fig. 2.11 Illustration of phase shift due to **a** scalar and **b–d** vector potentials

2.4.2 Phase Shift Due to Vector Potential

The phase shift due to a magnetic field is evaluated as follows. If the specimen thickness is uniform as shown in Fig. 2.11b, $\phi(x, y)$ can be expressed with the vector potential $\overrightarrow{A}$ related to magnetic flux density B:

$$\phi(x, y) = -\frac{e}{\hbar}\int \vec{A}(x, y, z)d\vec{z}, \tag{2.24}$$

where e and $\hbar$ are the elementary electric charge and Planck's constant divided by 2π, respectively. If the electron wave is assumed to have the same phase at points A and B above the specimen in Fig. 2.11b, the phase difference between points C and D below the specimen is given by

$$\begin{aligned}\phi(x_C, y_C) - \phi(x_D, y_D) &= -\frac{e}{\hbar}\int_A^C \vec{A}(x, y, z)\cdot d\vec{z} + \frac{e}{\hbar}\int_B^D \vec{A}(x, y, z)\cdot d\vec{z},\\ &= \frac{e}{\hbar}\oint_{ABDC} \vec{A}(x, y, s)d\vec{s}.\end{aligned} \tag{2.25}$$

With the vector potential in (1.8) and Stokes' theorem, the following relation can be obtained

$$\begin{aligned}\oint \vec{A}\cdot d\vec{s} &= \iint rot\vec{A}\cdot d\vec{S}\\ &= \iint \vec{B}\cdot d\vec{S}\\ &= \iint B_n\, dS,\end{aligned} \tag{2.26}$$

where B_n is the component of $\vec{B}$ normal to area dS. Thus, we obtain

$$\phi(x_C, y_C) - \phi(x_D, y_D) = \frac{\mathrm{e}}{\hbar}\iint B_n dS$$

$$= \frac{e}{\hbar}\Phi, \tag{2.27}$$

where Φ is the magnetic flux passing through and normal to the area *ABDC*. If the phase difference between $\phi(x_C, y_C)$ and $\phi(x_D, y_D)$ is 2π, we have

$$\frac{e}{\hbar}\Phi = 2\pi, \tag{2.28}$$

and then

$$\phi = \frac{h}{e} = 4.1 \times 10^{-15}\ \mathrm{Wb}. \tag{2.29}$$

This value is twice the flux quantum, i.e., *h*/2*e*. Further, we have the relationship between the magnetic flux inside the specimen and width l of the white (or black) lines in the reconstructed phase image $(\cos(\phi(x, y)))$ corresponding to a 2π phase difference,

$$\Phi = l\,t\,B_n = h/e. \tag{2.30}$$

From (2.30), the relationship between l and B_n is given by

$$l = \frac{h}{etB_n}. \tag{2.31}$$

Different from the scalar potential, the phase shift appears in (2.27) for a constant specimen thickness. Even if the electron beam does not pass through the magnetic field $\vec{B}$, there is a difference in phase shifts between positions C and D. This is because $\vec{B}$ exists between the two beam paths: A–C and B–D, or this difference results from the existence of vector potential $\vec{A}$, shown in Fig. 2.11c. This is the so-called Aharonov-Bohm effect [21].

Tonomura and his colleagues extended the electron holography system using a toroidal ferromagnet covered with a superconductor (Fig. 2.11d). Using this extended system, they were able to clearly prove the physical reality of vector potentials [22].

2.4.3 Separation of Magnetic Information from Electric Information

The phase information of the electric and magnetic fields can be separated using the "time-reversal operation of an electron beam" proposed by Tonomura et al. [23]. A pair of electron holograms are obtained in the same area, but the foil is reversed against the incident electron beam. Note that the sign of the phase depends on the direction of the incident electron beam for the magnetic field, whereas it is independent of the electric field (2.22). For separating the phase information of the electric field from that of the magnetic field, techniques of applying an external magnetic field and setting the temperature to various values around the magnetic phase transformation temperature can be also utilized.

Here, we describe the separation of magnetic and electric information by using the time-reversal operation of an electron beam for a stainless-steel sheet specimen containing localized magnetic domains. Figure 2.12a shows the reconstructed phase image in which the localized magnetic domain is on the left. The reconstructed phase image in Fig. 2.12b was obtained by turning the specimen upside down. Figure 2.12c, d correspond to the reconstructed phase images showing the electric and magnetic fields, respectively. The series of image treatments are shown in Fig. 2.12e. The reconstructed phase image in the geometrical condition of the wedge shape specimen (A) corresponds to the image in (a). The geometrical condition obtained by turning the specimen upside down is shown in B. Image B' was obtained by flipping image B, resulting in the same geometrical condition as A. Reconstructed phase image (c) showing the electric field can be obtained by calculating (A + B′)/2 while reconstructed phase image (d) corresponding to the magnetic field can be obtained by calculating (A−B′)/2.

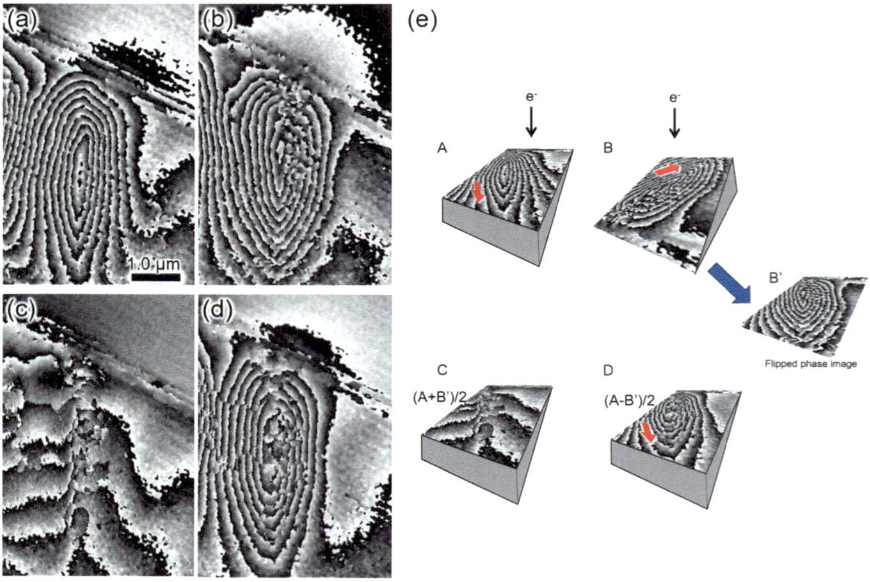

Fig. 2.12 **a** Reconstructed phase image of stainless-steel specimen. **b** Reconstructed phase image obtained with specimen turned upside down. **c**, **d** Phase images corresponding respectively to electric and magnetic phase shifts obtained from (**a**) and (**b**). **e** Processes applied to phase images to separate electric (C) and magnetic (D) phase shifts corresponding to (**c**) and (**d**), respectively. A and B' correspond to phase images (**a**) and (**b**), respectively. Red arrows in **e** indicate direction of magnetic flux. Phase images were obtained in collaboration with Dr. Taniyama

2.5 Simulation of Interference Fringes

2.5.1 Formation of Interference Fringes

According to Komrska [24], we explain the fundamental principles of interference fringe formation in several steps [6]. The geometrical condition in the wave-optical treatment is shown in Fig. 2.13. The electron source P_0 is located far from the biprism wire. The biprism wire field drops quickly to a few micrometers as shown in Fig. 2.14. Thus, the spherical wave $\psi(P_0)$ emitted from P_0 can be treated as a plane wave at the biprism position. The plane wave becomes a wave function $\psi(M)$ that is affected by the electric field of the biprism and generates wave function $\psi(P)$ at the observation point as explained in the following steps [6]. Our objective is to obtain wave function $\psi(M)$ at the biprism position from the wave function $\psi(P_0)$ at the electron source and then to express the wave function at the observation point using $\psi(M)$. The coordinate system at each position is shown in Fig. 2.15.

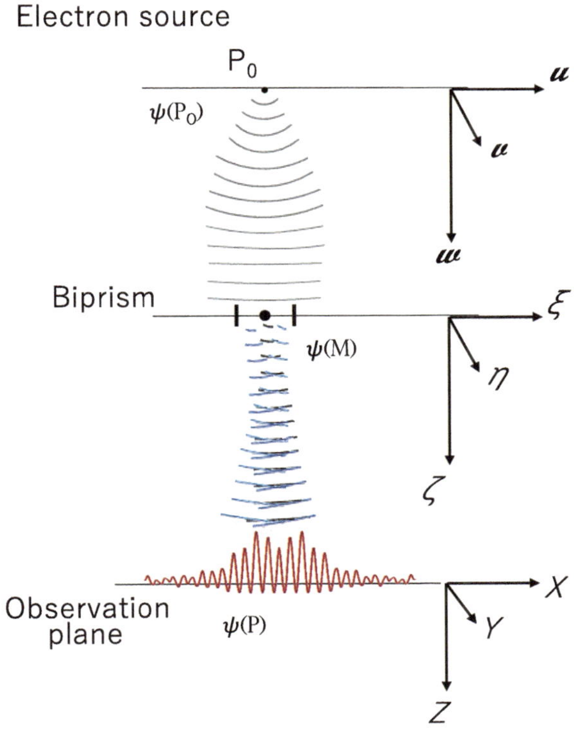

Fig. 2.13 Illustration of interference fringe formation based on wave-optical treatment

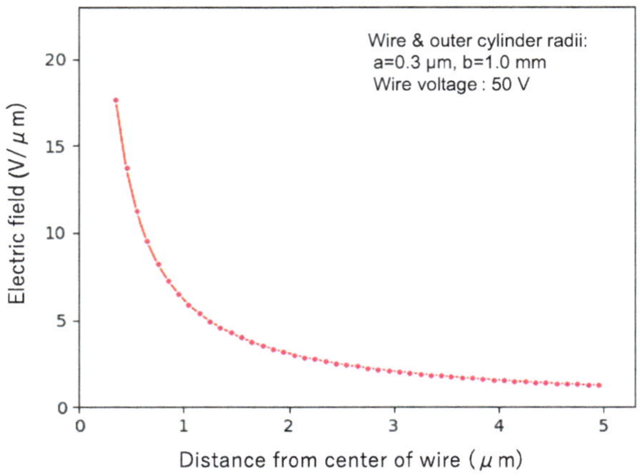

Fig. 2.14 Electric potential as function of distance from electron source to biprism wire

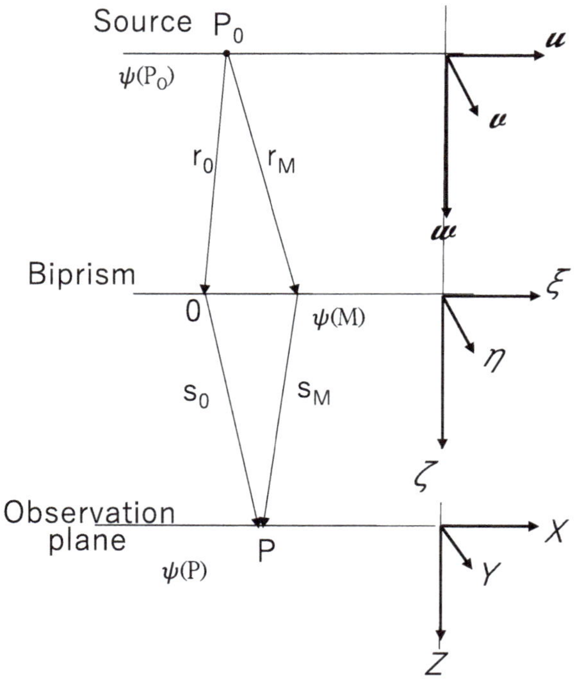

Fig. 2.15 Illustration showing ray paths from electron source to observation plane

(1) Wave Function at Wire Plane

If $\psi_U(M)$ is the solution to Schrödinger's equation of the wave function affected by the electric field of the wire, using a transmission function *f(M)*, $\psi(M)$ is given as

$$\psi(M) = \psi_U(M) \cdot f(M), \tag{2.32}$$

where *f(M)* represents only the amplitude and $\psi_U(M)$ is given with path (*S*) as a function of $\vec{r}$ and *E*:

$$\psi_U(\vec{r}) = A(\vec{r})\exp\left[ikS(\vec{r}, E)\right]. \tag{2.33}$$

Since $\psi_U(\vec{r})$ satisfies Schrödinger's equation, $S(\vec{r}, E)$ is expressed as

$$S(\vec{r}, E) = l - \frac{1}{2E}\int_l U(\vec{r})dl. \tag{2.34}$$

When we set

$$A(\vec{r}) = |\psi_0(\vec{r})|, \tag{2.35}$$

we have the relation, i.e.,

$$\psi_U(\vec{r}) = |\psi_0(\vec{r})|\exp\left\{ik\vec{r}\left[1 - \frac{1}{2E}\int_0^1 U(P_0 + \vec{r}t)dt\right]\right\}. \tag{2.36}$$

(2) Green's Integral Theorem

Wave function ψ(P) at the observation point is obtained by calculating the wave motion generated at the observation point on the basis of the waves from the surrounding wave sources. We discuss the theorem using the spaces and interfaces defined in Fig. 2.16: biprism surface plane S_1 contains the wire, and sphere *S'* surrounds observation point *P*. Point *P* is surrounded by volume V, which is surrounded by these connection surfaces: $S = S_1 U S_2 U S'$.On the basis of Gauss' theorem, wave functions ψ and ψ_1 satisfy

$$\iiint_V \left(\psi\nabla^2\psi_1 - \psi_1\nabla^2\psi\right)dV = \iint_S (\psi\nabla\psi_1 - \psi_1\nabla\psi)\vec{n}dS, \tag{2.37}$$

where $\vec{n}$ is a unit vector perpendicular to the surface. Since ψ and ψ_1 are solutions to Schrödinger's equation, the term in the volume integral is zero. By applying Green's function G to ψ_1, we get

$$\iint_{S_1 \cup S_1} (\psi \nabla G - G \nabla \psi) \vec{n} dS + \iint_{S'} (\psi \nabla G - G \nabla \psi) \vec{n} dS = 0, \tag{2.38}$$

According to the treatment by Sommerfeld [25]. The integral of surface *S'* including observation point *P* is the extremum of the surface area surrounding *P* when *S'* approaches 0. In the limit of the integral over sphere *S'* in (2.38), the following relation is obtained, i.e., we get

$$\lim_{s \to 0} \iint_{S'} (\psi \nabla G - G \nabla \psi) \vec{n} dS = \psi(P) \int_0^{2\pi} \int_0^{\pi} \sin \vartheta d\vartheta d\chi = 4\pi \psi(P). \tag{2.39}$$

Substituting this equation into (2.38), we obtain

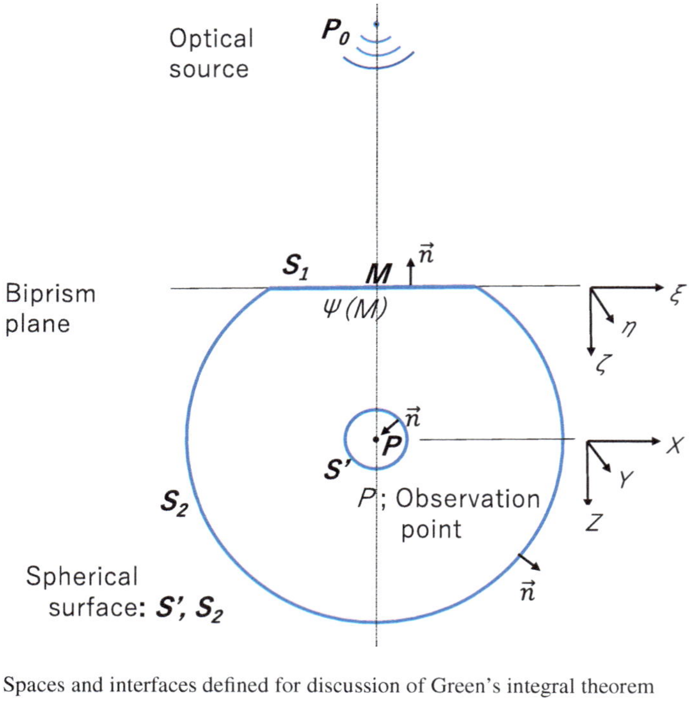

Fig. 2.16 Spaces and interfaces defined for discussion of Green's integral theorem

$$\begin{aligned}\psi(P) &= -\tfrac{1}{4\pi}\iint\limits_{S_1\cup S_2}(\psi\nabla G - G\nabla\psi)\vec{n}dS \\ &= -\tfrac{1}{4\pi}\iint\limits_{S_1}\psi\nabla G\vec{n}dS - \tfrac{1}{4\pi}\iint\limits_{S_2}(\psi\nabla G - G\nabla\psi)\vec{n}dS,\end{aligned} \tag{2.40}$$

where we used $G(P,M) = 0$ at points $Q \neq P$ of the half-space ($\varkappa$,P). When the radius (s) of S_2 approaches infinity, the integration of S_2 approaches 0:

$$\lim_{s\to\infty}\frac{1}{4\pi}\iint\limits_{S_2}(\psi\nabla G - G\nabla\psi)\vec{n}dS = 0. \tag{2.41}$$

Thus, only the integral for S_1 remains as

$$\psi(P) = -\frac{1}{4\pi}\iint\limits_{S_1}\psi(M)\nabla G(P,M)\vec{n}dS. \tag{2.42}$$

(3) Explicit Form of Green's Function

Green's function $G\,(P, M)$ in (2.42) corresponds to the propagation of the wave from the wire plane to the observation plane. By means of "the method of images" [25], Green's function $G\,(P, Q)$ for points M and P shown in Fig. 2.17 is given by

$$\begin{aligned}G(P,Q) = &\tfrac{1}{s_Q}\exp\left\{iks_Q\left[1 - \tfrac{1}{2E}\int_0^1 U_g(P + \vec{s}_Q t)dt\right]\right\} \\ &-\tfrac{1}{s'_Q}\exp\left\{iks'_Q\left[1 - \tfrac{1}{2E}\int_0^1 U_g\left(P' + \vec{s}'_Q t\right)dt\right]\right\}.\end{aligned} \tag{2.43}$$

$$(U_g(\zeta,\eta,\varsigma) = U(\zeta,\eta,\varsigma) \quad \text{if } \varsigma \geqq 0, \quad U_g(\zeta,\eta,\varsigma) = U(\zeta,\eta,-\varsigma) \quad \text{if } \varsigma \leqq 0)$$

In (2.43), the first term corresponds to the spherical wave from point Q to point P. The second term represents the spherical wave from point P', which is the point symmetric to observation point across wire plane, to point Q. Further, under the conditions $1/(k\cdot s_M) \ll 1$ and $U/2E \ll 1$, we obtain

$$\nabla G\cdot\vec{n} = 2ik\cdot\exp\left\{iks_M\left[1 - \frac{1}{2E}\int_0^1 U(P + s_M t)dt\right]\right\}\frac{s_M\cdot\vec{n}}{s_M^2}, \tag{2.44}$$

where U is the potential along the electron path and s_M is the path from the biprism to the observation plane.

(4) Intensity Distribution of Interference Fringes

The wave function at observation plane $\psi(P)$ given in (2.42) can now be given by inserting $\psi(M)$ in (2.32) and ∇G in (2.44):

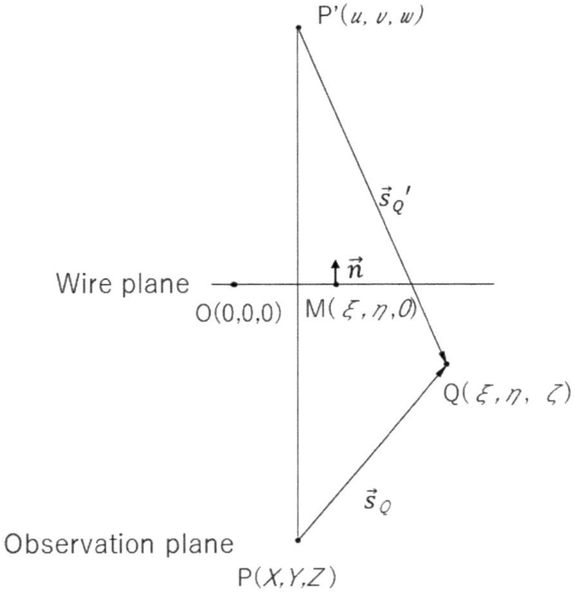

Fig. 2.17 Geometrical configuration for discussion of Green's half-space function

$$\psi(P) = D \iint_S f(M) \exp\{i[\phi(M) + \phi_U(M)]\} dS_M, \tag{2.45}$$

where the amplitude and phase are given by the following equations, i.e.,

$$D = -\frac{ikA_0}{2\pi r_0 s_0} \exp[ik(r_0 + s_0)] \cdot \exp\left[-\frac{ik}{2E} \int_0^1 [r_0 U(P_0 + r_0 t) + s_0 U(P + s_0 t)] dt\right] \tag{2.46}$$

$$\left(|\psi_0(r_M)| = \frac{A_0}{r_M};\ A_0 : \text{amplitude of source} \right)$$

$$\phi(M) = k[r_M + s_M - (r_0 + s_0)] \tag{2.47}$$

$$\phi_U(M) = -\frac{k}{2E} \int_0^1 \left[r_M U(P_0 + \vec{r}_M t) + s_M U(P + \vec{s}_M t) - r_0 U(P_0 + \vec{r}_0 t) - s_0 U(P + \vec{s}_0 t) \right] dt \tag{2.48}$$

These three equations are used to evaluate interference fringe intensity.

For the cylindrical capacitor potential shown in Fig. 2.18, $\psi(P)$ is given as

$$\psi(P) = D_2 \left[\int_{-\xi_{max}}^{-a} + \int_a^{\xi_{max}} \right] \exp\{i[\phi(\xi) + \phi_U(\xi)]\} d\xi, \tag{2.49}$$

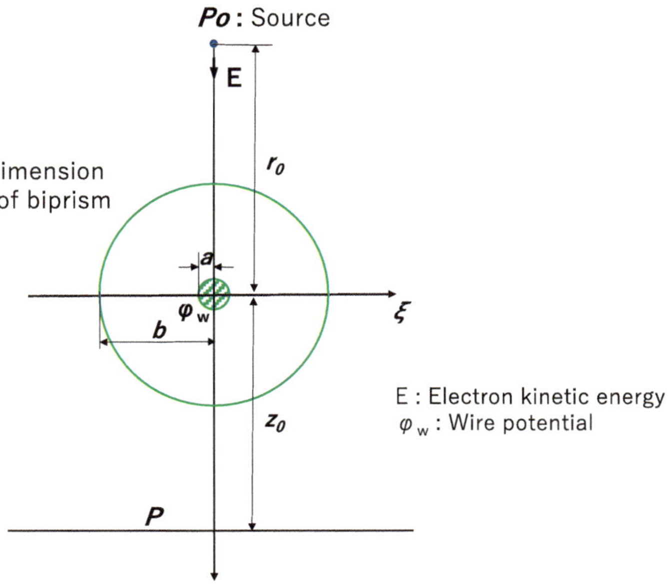

Fig. 2.18 Geometry of cylindrical capacitor for calculation of interference fringe intensity

where *D2* and the exponential term are given as follows:

$$D_2 = \frac{\exp(-i\pi/4)}{\pi}\frac{A_0}{(r_0 z)^{1/2}}\exp\left[ik\left(R - b\frac{|e|\varphi_w}{\mathrm{In}(a/b)}\frac{1}{E}\right)\right] \tag{2.50}$$

$$\phi(\xi) + \phi_U(\xi) = k\left\{\frac{1}{2}\left(\frac{1}{r_0} + \frac{1}{z_0}\right)\left(\xi - \frac{x r_0}{r_0 + z_0}\right)^2 + \frac{|e|\varphi_w}{\mathrm{In}(a/b)}\frac{1}{2E}\cdot \right.$$
$$\left. \xi\left[\left(\frac{\xi}{r_0} + \frac{\xi - x}{z_0}\right)\mathrm{In}\frac{|\xi|}{b} + \pi\cdot\mathrm{sgn}\xi - \frac{\xi}{b}\left(1 + \frac{1}{12}\frac{\xi^2}{b^2}\right)\right]\right\}. \tag{2.51}$$

Finally the relative intensity distribution of the interference fringes at the observation plane is given from (2.49), i.e.,

$$I(X) = \frac{\psi\psi^*}{\psi_0\psi_0^*} = \frac{k}{2\pi}\left(\frac{1}{r_0} + \frac{1}{z_0}\right)\left|\left[\int_{-\xi_{max}}^{-a} + \int_{a}^{\xi_{max}}\right]\exp\{i[\phi(\xi) + \phi_U(\xi)]\}d\xi\right|^2. \tag{2.52}$$

The formulations derived so far in this section are next used to characterize the features of the interference fringe distribution such as the stationary points, interference region, and interference fringe spacing. Furthermore, they are used to simulate the detailed interference fringe distribution, which is compared with the observed fringe distribution in Sect. 2.5.2.

(5) Stationary Points and Spacing of Interference Fringes

If we differentiate (2.51) by coordinate ξ of the wire plane and set it to 0,

$$\frac{d}{d\xi}\{\phi(\xi)+\phi_U(\xi)\}=0. \tag{2.53}$$

From this equation, we can obtain points ξ_1 and ξ_2 on the wire plane:

$$\xi_1=\frac{r_0 x}{r_0+z_0}-\pi\frac{r_0 z_0}{r_0+z_0}\frac{e\varphi_w}{\ln(a/b)}\frac{1}{2E} \tag{2.54}$$

$$\xi_2=\frac{r_0 x}{r_0+z_0}+\pi\frac{r_0 z_0}{r_0+z_0}\frac{e\varphi_w}{\ln(a/b)}\frac{1}{2E}. \tag{2.55}$$

$$(\ln(a/b)\ <\ 0)$$

These equations show that there are two stable paths with a stationary phase. It is noted that the contributions of these two paths mainly result in the formation of interference fringes. When ξ_1 or ξ_2 reaches the end of the wire, the fringe disappears because the wave of the stable phase forming the fringe is screened off by the wire. The points on the observation plane in this case are X_A and X_B:

$$X_{A,B}=\pm\left[\pi z_0\frac{e\varphi_w}{\ln(a/b)}\frac{1}{2E}+\frac{a}{r_0}(r_0+z_0)\right]. \tag{2.56}$$

The width W of the interference region is obtained by subtracting X_A from X_B, with $\xi_1=a$ and $\xi_2=-a$ at the two ends of the wire, respectively,

$$W=X_B-X_A=-2\left[\pi z_0\frac{e\varphi_w}{\ln(a/b)}\frac{1}{2E}+\frac{a}{r_0}(r_0+z_0)\right]. \tag{2.57}$$

The wave function at the points on the observation plane is mostly contributed to by the waves from stationary points *M1* and *M2*, and it reaches maximum when the phase difference between points *M1* and *M2* is 0, ± π, ± 2πs,.... Thus, we can define

$$X_n=-n\frac{r_0+z_0}{r_0}\cdot\ln(a/b)\cdot\frac{\lambda E}{2\pi e\varphi_w}. \tag{2.58}$$

$$(\ln(a/b)<0)$$

The fringe interval d is given by

$$d=X_{n+2}-X_n=-\frac{(r_0+z_0)}{r_0}\cdot\ln(a/b)\cdot\frac{\lambda E}{\pi e\varphi_w}. \tag{2.59}$$

2.5.2 Simulation of Interference Fringes

Figure 2.19a shows interference fringes observed using a 300-kV electron microscope equipped with a CCD camera. Figure 2.19b shows the intensity profile of the fringes in band region A–B in (a). The normalized intensity profile of the interference fringes evaluated from (2.52) is shown in (d). Table 2.3 indicates the parameters and values used in the simulation. The fringe spacing and intensity profile of the observed interference fringes with the Fresnel fringes at both ends of the intensity profile (Fig. 2.19b) are reproduced in the simulation.

To take into account the contribution of the effective coherence lengths caused by the illumination angle and the energy spread of the incident electrons, the simulated interference fringes were broadened by superimposing slightly shifted profiles as shown in Fig. 2.19d. The resulting profile in (c) is in good agreement with the observed one in (b).

2.6 Lorentz Microscopy

In this section, we describe another technique using an electron microscope that can be used to visualize domain structures and/or domain walls. Since the principles of this imaging technique are understood in terms of Lorentz force on the incident electrons, the imaging technique is called "Lorentz microscopy."

Electron holography requires a highly coherent condition and thus the illumination angle should be small, which means that collimated illumination is necessary. Also, the basic principles of the imaging process in electron holography are formulated via quantum mechanics accordingly.

On the other hand, Lorentz microscopy observation is understood through the classical particle picture of incident electrons. The illumination angle of the incident electron beam in Lorentz microscopy can be larger than that in electron holography and a more intense electron beam can be used. Thus, this technique is thus useful for dynamic observation of electromagnetic fields with a video system accordingly [9, 26].

Table 2.3 Parameters used for simulation of interference fringes

Parameter	Symbol	Value used for simulation
Energy of electrons	E	300 keV
Wave number	k	1.95×10^7 mm^{-1}
Source to wire	r_0	200 mm
Wire to observation plane	z_0	71 mm
Wire diameter	$2a$	5×10^{-4} mm
Outer cylinder diameter	$2b$	1 mm
Wire potential	φ_w	35 V

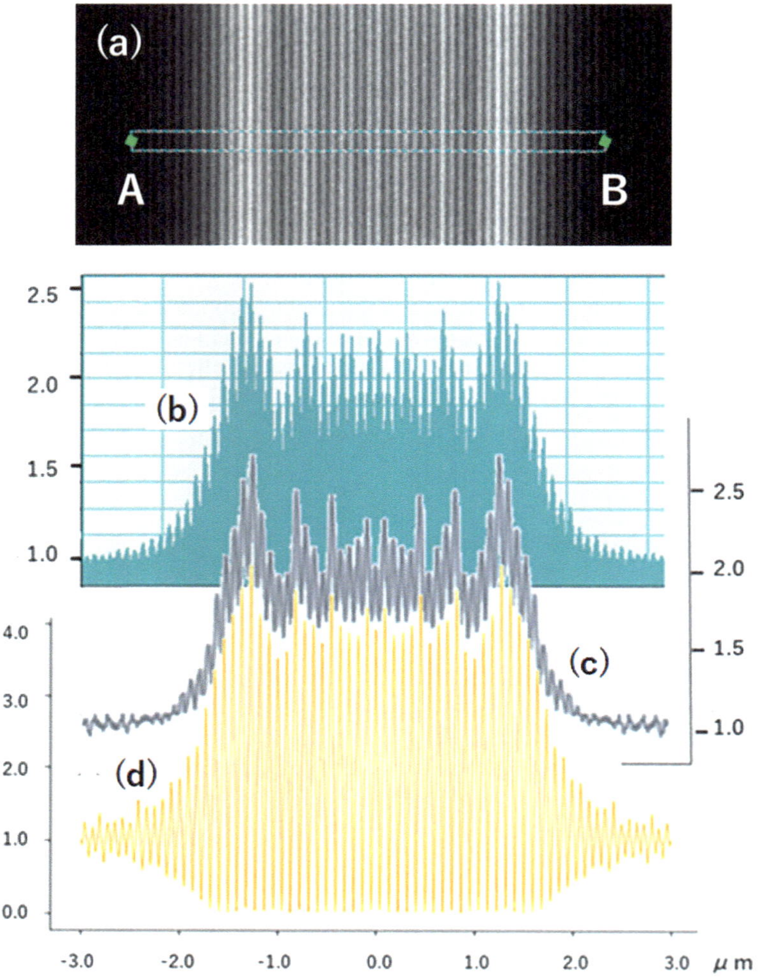

Fig. 2.19 Comparison of profiles of observed and simulated interference fringes. **a** Intensity distribution of interference fringes observed at wire potential of 35 V and accelerating voltage of 300 kV. **b** Intensity profile of observed interference fringes averaged in vertical direction for region A–B in (**a**). **c** Intensity profile of simulated interference fringes obtained by taking into account contribution of illumination angle and energy spread of incident electrons. **d** Normalized intensity profile of interference fringes simulated using parameter values in Table 2.3

2.6.1 Fresnel Mode (Defocusing Mode)

Lorentz force $\vec{F}$ for an electron passing through a specimen with magnetic flux density $\vec{B}$ (T) is given by

$$\vec{F} = -e\left(\vec{v} \times \vec{B}\right), \tag{2.60}$$

where $\vec{v}$ is a vector of the electron velocity. The direction of the force expressed by the vector product in (2.60) is given by Fleming's left-hand rule, and the trajectory of the incident electrons is as shown in Fig. 2.20a, where the 180° domains are magnetized perpendicularly to the plane of the diagram and in opposite directions in alternate domains in (b). The electron beam is thus deflected in opposite directions in adjacent domains, forming alternating deficiency and excess of electrons in the region below the lower specimen surface. If we observe the specimen in the over-focus condition, increases and decreases in the electron intensity are observed periodically at domain wall positions in Fig. 2.20c. On the other hand, if we observe the specimen in the under-focus condition, the image contrast is reversed in Fig. 2.20d.

Figure 2.21a, b show Lorentz microscopy images of an as-quenched $Fe_{73.5}Cu_1Nb_3Si_{13.5}B_9$ specimen observed in under- and over-focus conditions, respectively. These images show magnetic domain boundaries as white and black

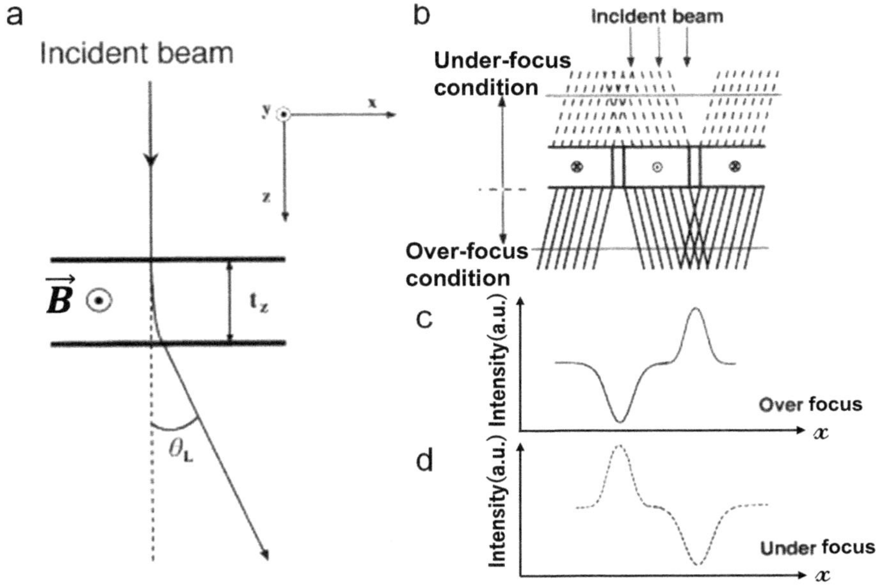

Fig. 2.20 **a** Trajectory of electrons passing through magnetized specimen. **b** Schematic of Lorentz microscopy in Fresnel mode. Intensity distributions for **c** over- and **d** under-focus conditions

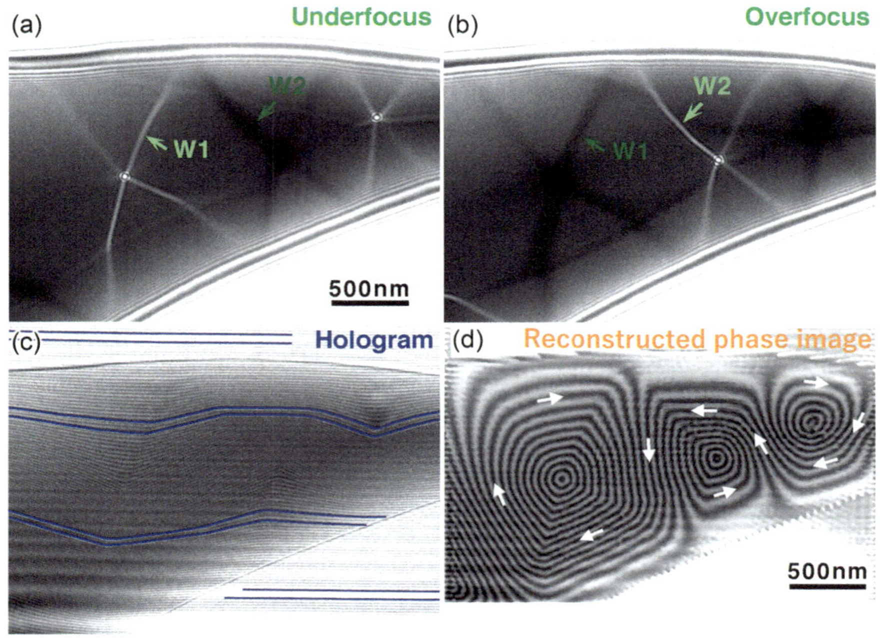

Fig. 2.21 Lorentz microscopy images of as-quenched $Fe_{73.5}Cu_1Nb_3Si_{13.5}B_9$ specimen observed in **a** under- and **b** over-focus conditions. **c** Electron hologram of specimen. **d** Reconstructed phase image of specimen (arrows indicate direction of lines of magnetic flux). Permission obtained from Elsevier [26]

bands as indicated by W1 and W2. In Fig. 2.21c, an electron hologram of the same area is shown. It is noted that the interference fringes switch direction from time to time. In the reconstructed phase image represented by $\cos(\phi(x, y))$ in (d), the densities and directions (arrows) of the white lines correspond to the densities and directions of the magnetic flux lines, respectively.

In Fig. 2.21d, smooth closure domains are clearly seen through the lines of magnetic flux. It is seen that the magnetic flux lines are parallel to the specimen edges, thereby eliminating the surface magnetic charge. It is also noted that the domain walls observed in (a) correspond to the boundaries in (d), in which the lines turn about 90°. The specimen thickness gradually increases from the edge to the middle part as the thickness increases, and the spacing between the lines decreases, corresponding to the larger magnetic flux projected along the incident electron beam accordingly.

2.6.2 Foucault Mode (In-Focus Mode)

Figure 2.22 shows the electron diffraction pattern obtained from an area containing a domain wall separating two magnetic domains. It is seen that the diffraction spots split and form two spots. This splitting is observed in the electron diffraction pattern, for both the transmitted and diffracted beams. The Lorentz force on an electron passing through a magnetized specimen with magnetization $\overrightarrow{I_s}$ (A/m) is given by

$$\vec{F} = -e\mu_0\left(\vec{v} \times \overrightarrow{I_s}\right), \tag{2.61}$$

where μ_0 (H/m) is the permeability in vacuum. Here, the direction of the magnetization in a specimen is assumed to be normal to the plane of the diagram, i.e., the y-direction, as shown in Fig. 2.20. Under this condition, the electron is deflected from the straight trajectory due to magnetization $\overrightarrow{I_s}$, but its energy is unchanged. From (2.61), the Lorentz force on the incident electron, which is parallel to the x-direction, is expressed as

$$\frac{\mathrm{d}p_x}{\mathrm{d}t} = m'\frac{\mathrm{d}^2x}{\mathrm{d}t^2} = ev\mu_0, \tag{2.62}$$

where m' is the electron mass at velocity v:

$$m' = m/\left(1 - \frac{v^2}{c^2}\right)^{\frac{1}{2}}, \tag{2.63}$$

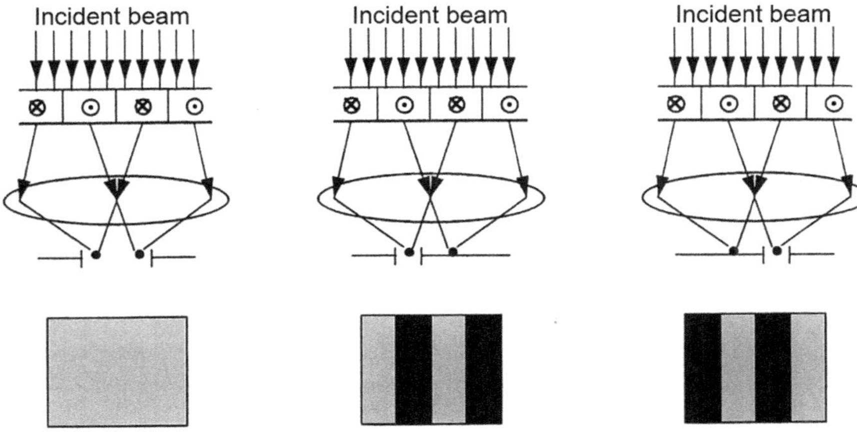

Fig. 2.22 Schematic illustration of Lorentz microscopy in Foucault mode

with m being the rest mass. In (2.62), $\mathrm{d}p_x/\mathrm{d}t$ is the x-component of the Lorentz force with $p_x = m'(\mathrm{d}x/\mathrm{d}t) = et_z\mu_0 I_y$, because the electron's velocity can be treated as a constant. For the magnitude of the magnetization given by $I_y = I_s$ and $I_x = I_z = 0$, deflection angle θ_L is given by

$$\theta_L = \frac{\mathrm{d}x}{\mathrm{d}z} = \frac{\mathrm{d}x}{\mathrm{d}t}\frac{\mathrm{d}t}{\mathrm{d}z} = \frac{et_z}{m'v}\mu_0 I_y. \tag{2.64}$$

The angle of deflection due to the Lorentz force is small compared with the Bragg angle. For example, deflection angle θ_L for $Nd_2Fe_{14}B$ with a thickness of 100 nm and $B = \mu_0 I_s = 1.6$ T is 3×10^{-5} rad for 1250-keV electrons. The electron diffraction pattern in Fig. 2.23a, obtained for a sintered $Nd_2Fe_{14}B$ magnet, shows the splitting of each spot. The separation of the spots was ~5×10^{-5} rad, which is consistent with the estimation. By displacing the objective aperture in the back focal plane, it is possible to exclude one of the two transmitted beams. As a result, alternate magnetic domains appear periodically as dark and bright regions, as shown in (b) and (c).

Figure 2.24a, b show Lorentz microscopy images of a step-aged Sm-Co magnet observed in Fresnel and Foucault modes, respectively. The image in (a) shows only serrated domain walls, indicating pinning at the cell boundaries. On the other hand, the image in (b) shows not only the magnetic domain but also the microstructure. The microstructure of this magnet consists of 1:5 H and 2:17 R phases, where the 1:5 H phases of the cell boundaries act as attractive pinning centers for the magnetic domain walls in the step-aged magnet, as illustrated in (c). It is seen that domain wall pinning increases greatly with an increase in the domain wall energy gradient between the 1:5 H and 2:17 R phases owing to the chemical partitioning of the additives of Cu and Fe atoms by step aging [9, 27].

A modified observation mode was developed for the Foucault mode [28] by eliminating half the diffraction pattern by using a phase-shifting aperture, and enabling the interference contrast in the domain to be obtained. The aperture itself is one of three types, i.e., opaque half-plane, phase-shifting half-plane, or phase-shifting small

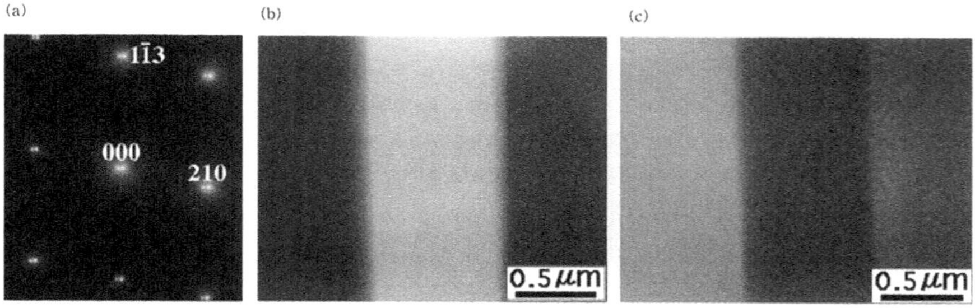

Fig. 2.23 **a** Electron diffraction pattern of sintered $Nd_2Fe_{14}B$ magnet. (b, c) Typical Lorentz microscopy images of magnetic domains observed in Foucault mode when objective aperture eliminated one of split transmitted beams. Permission obtained from Elsevier [26]

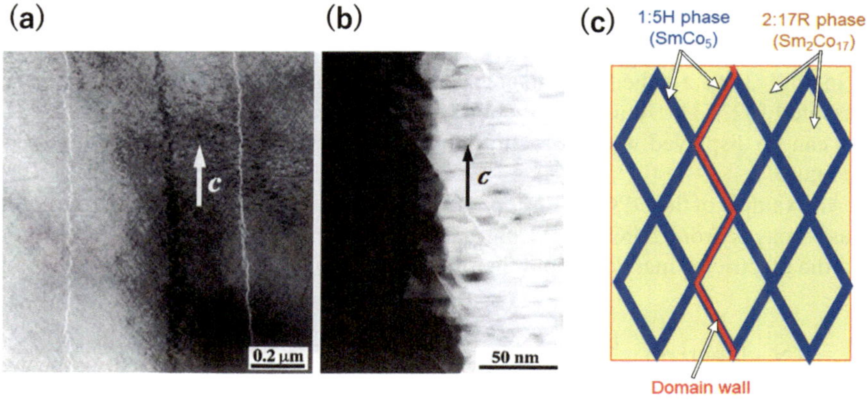

Fig. 2.24 Lorentz microscopy images of step-aged Sm-Co magnet observed in **a** Fresnel mode and **b** Foucault mode. Arrow in (**a**) indicates [001] direction. **c** Schematic illustration of geometrical relationship between domain wall (along $SmCo_5$ phase) and specimen microstructure. Permission obtained from Elsevier [26]

hole. The mode is called the "coherent Foucault mode." If the edge of the phase-shifting aperture is put at the center of a transmitted beam, interference contrast can be observed in all directions. A coherent Foucault image is obtained in real time without any postprocessing, so this imaging technique is useful for in situ experiments [29].

2.6.3 Lorentz Microscopy Using Scanning Transmission Electron Microscope

Differential phase contrast (DPC) Lorentz microscopy, introduced by Chapman et al. [30], provides detailed imaging of the magnetization distribution in magnetic materials like electron holography described in Sect. 2.1. The principles of DPC Lorentz scanning TEM (STEM) are explained as follows [31].

In scanning mode, the electron beam is converged by an objective pre-field lens and focused on the magnetic specimen. At each point, the disc pattern of the transmitted beam at the back focal plane of the objective lens is slightly shifted from the center of the ray axis because of the beam's deflection due to the Lorentz force. Figure 2.25 shows how the disc on the objective back focal plane is magnified at the split detector with intermediate and projector lenses. When the convergent beam passes through the magnetic domain, as shown in (a), the deflection disc is shifted to the right of the detector, and the signal detected by detector (B + C) is stronger than that detected by detector (A + D), and vice versa, as shown in (b, c). If the signal detected by (B + C) is subtracted from that detected by (A + D), the signal component due to the phase shift is enhanced while the component not due to the phase shift vanishes since the latter component is detected equally by both detectors. Thus, the magnetization direction can be detected accordingly.

In Fig. 2.26, a schematic diagram of a quadrant-detector imaging system is presented [32, 33]. By adding or subtracting each signal, domains in every direction can be displayed without rotating the specimen, as shown on the right side of the figure.

An example of the DPC-STEM operation screen is shown in Fig. 2.27. Differential phase images from DPC-STEM are calculated in situ at the time of acquisition, and the electric or magnetic field distribution is displayed on the operation screen.

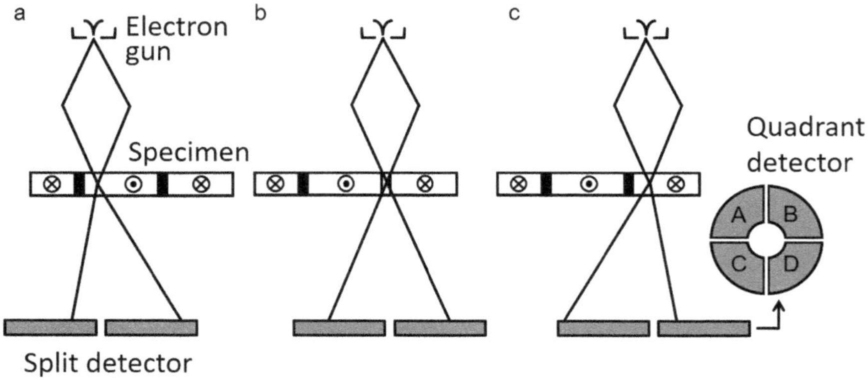

Fig. 2.25 Geometric configuration of electron gun and quadrant detector in DPC Lorentz microscopy

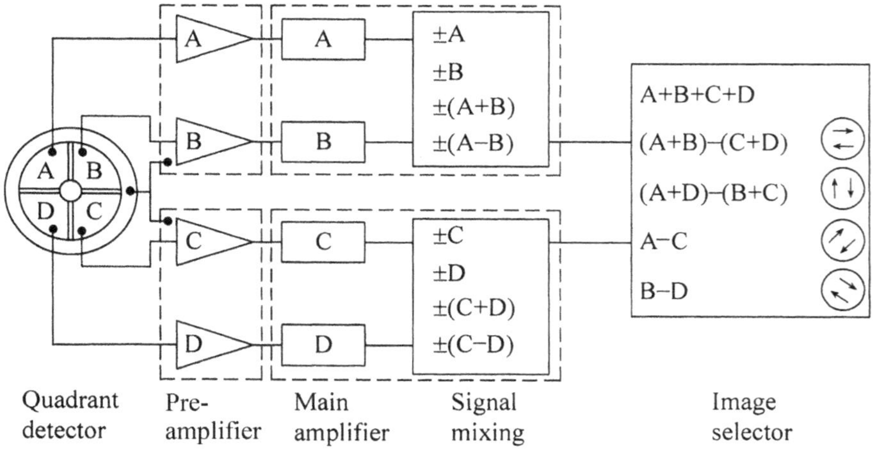

Fig. 2.26 Geometric configuration of quadrant detector and image processing system in DPC Lorentz STEM

However, the magnetic and electric field distributions are not separated, so if it is necessary to separate them, it must be done after data acquisition, as is the case with electron holography (see Fig. 2.12). Figure 2.27 shows the operation screen during DPC-STEM observation of the magnetic field distribution for Mn-Zn ferrite. Figure 2.27a, b are maps of the x- and y-components of the magnetic field in grayscale, respectively; (c) shows the goniometer information, where the TX entry represents the x-tilt value of the specimen; (d) is a color map of the magnetic field calculated on the basis of the information in (a) and (b); (e) is an image of the divergence of the magnetic field, which is similar in contrast to the Lorentz microscope image obtained for Fresnel mode. A video of DPC-STEM observation is presented in Fig. 2.28.

DPC Lorentz STEM with a probe size of less than 0.1 nm has been used to clarify the electric field between atomic columns [34]. It has been used to determine the three-dimensional electric field at an electron nano-emitter [35].

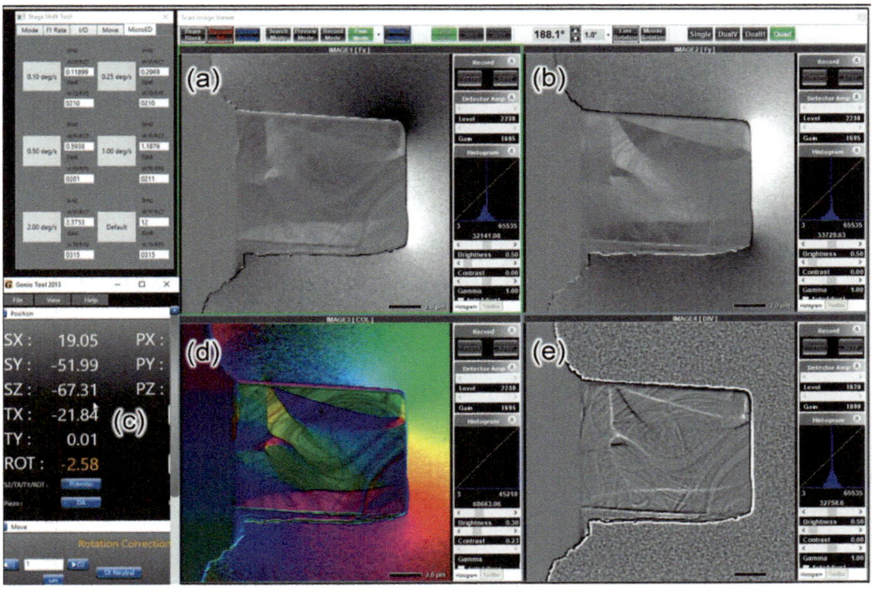

Fig. 2.27 Magnetic domain image of Mn-Zn ferrite captured by DPC Lorentz STEM: **a** x-component of magnetic field; **b** y-component of magnetic field; **c** goniometer information (TX entry represents x-tilt value of specimen); **d** color map of magnetic field; **e** image of magnetic field divergence

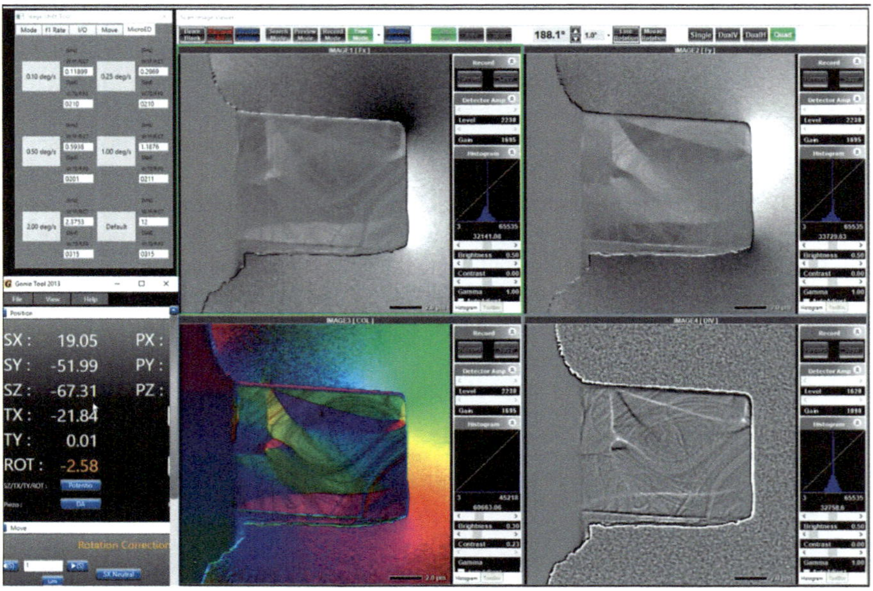

Fig. 2.28 Video: Magnetic domain image of Mn-Zn ferrite captured by DPC Lorentz STEM. Movie was prepared in collaboration with Dr. Yasuhara

2.6.4 Phase Reconstruction Using Transport-Of-Intensity Equation

By using the transport-of-intensity equation (TIE), the phase information of an electron beam can be obtained from a series of defocused Lorentz microscopy images captured in Fresnel mode [36–41]. TIE describes the internal relationship between the intensity and phase distribution when a wave is propagating in the space. The equation is given as

$$\frac{2\pi}{\lambda}\frac{\partial}{\partial z}I(x, y, z) = -\nabla_{xy} \cdot \left(I(x, y, z)\nabla_{\mathrm{xy}}\phi(x, y, z)\right), \tag{2.65}$$

where $I(x, y, z)$ is the intensity, ∇_{xy} is the differential operator in the x–y plane, and $\phi(x, y, z)$ is the phase shift. Paganin and Nugent suggested solving (2.65) by using a Poisson equation [42]. Solving the phase shift requires partial differentiation of the beam intensity in the direction of propagation. This partial differentiation can be approximated via divided differences of Lorentz microscopy images observed in Fresnel mode. The phase shift image can be calculated from three Lorentz microscopy images for which the defocus values are symmetric around the in-focus value [43, 44].

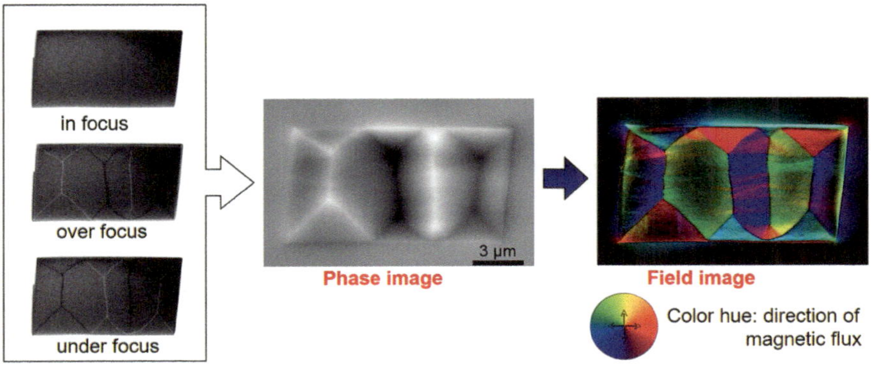

Fig. 2.29 Example of phase reconstruction using TIE equation for $Fe_{18.8}Co_{60}Cu_{0.6}Nb_{2.6}Si_9B_9$ nanocrystalline soft magnetic material. Permission obtained from Elsevier [26]

Figure 2.29 shows an example of this phase reconstruction using the TIE method. The specimen was an $Fe_{18.8}Co_{60}Cu_{0.6}Nb_{2.6}Si_9B_9$ nanocrystalline soft magnetic material thinned by focused ion beam milling. The left panel shows three Lorentz microscopy images. From these images, the phase image in the center panel was calculated using (2.65). The magnetic field image shown in the right panel, in which the magnetic flux is mapped by color hues, could be converted from the phase image because the specimen's thickness was almost uniform. Although the accuracy of the phase obtained using the TIE method is easily affected by low-frequency noise, this method is useful for analyzing magnetic domain structures because it does not require any special equipment for the electron microscope and does not impose any limitations on the field of view.

References

1. D. Gabor, Nature **161**, 777 (1948)
2. E.N. Leith, J. Upatnieks, J. Opt. Soc. Am. **53**, 1377–1381 (1963)
3. Z. Liu, D. Shindo, S. Ohnuma, H. Fujimori, J. Magn. Magn. Mater. **262**, 308 (2003)
4. G. Möllenstedt, The History of the Electron Biprism, in *Introduction to Electron Holography*. (Kluwer Academic, New York, 1999), p.1
5. J.M. Cowley, *Diffraction Physics*. (Elsevier New York, 1975)
6. D. Shindo, T. Tomita, *Material Characterization using Electron Holography* (Wiley-VCH, Weinheim, 2023)
7. W.O. Saxton, M.A. O'Keefe, D.J.H. Cockayne, M. Wilkens, Ultramicroscopy **13**, 349 (1984)
8. *International Tables for X-ray Crystallography*, Kynoch Press, Birmingham, England, vol. IV (1974)
9. D. Shindo, T. Oikawa, *Analytical Electron Microscopy for Materials Science* (Springer, Tokyo, 2002)
10. A. Tonomura, L. F. Allard, G. Pozzi, D. C. Joy, Y. A. Ono, *Electron Holography, Proceedings of the International Workshop on Electron Holography, Holiday Inn World's Fair, Knoxville, Tennessee, USA, August 29–31, 1994*, Elsevier (1995)

11. A. Tonomura, *Electron Holography*,2 nd ed., Springer, (1999)
12. H. Lichte, M. Lehmann, Reports. Prog. Phys. **71**, 16102 (2007)
13. D. Shindo, Y. Murakami, Microscopy **60**, S225 (2011)
14. E. Völkl, L.F. Allard, D.C. Joy, *Introduction to Electron Holography*, Springer Science & Business Media (2013)
15. T. Tanigaki, K. Harada, Y. Murakami, K. Niitsu, T. Akashi, Y. Takahashi, A. Sugawara, D. Shindo, J. Phys. D Appl. Phys. **49**, 244001 (2016)
16. J. Zweck, J. Phys. Condens. Matter **28**, 403001 (2016)
17. R.E. Dunin-Borkowski, A. Kovács, T. Kasama, M.R. McCartney, D.J. Smith, in *Springer Handb. Microsc.* (Eds.: P.W. Hawkes, J.C.H. Spence), Springer International Publishing, Cham, pp. 767–818 (2019)
18. D. Shindo, K. Hiraga, *High Resolution Electron Microscopy For Materials Science* (Springer, Tokyo, 1998)
19. H. Lichte, Ultramicroscopy **108**, 256 (2008)
20. D. Shindo, Y. Murakami, J. Phys. D Appl. Phys. **41**, 183002 (2008)
21. Y. Aharonov, D. Bohm, Significance of electromagnetic potentials in the quantum theory. Phys. Rev. **115**(3), 485 (1959)
22. A. Tonomura, N. Osakabe, T. Matsuda, T. Kawasaki, J. Endo, S. Yano, H. Yamada, Phys. Rev. Lett. **56**(8), 792 (1986)
23. A. Tonomura, T. Matsuda, J. Endo, T. Arii, K. Mihama, Phys. Rev. B Condens. Matter **34**, 3397 (1986)
24. J. Komrska, Scalar diffraction theory in electron optics. Adv. Electron. Electron Phys. **30**, 139 (1971)
25. A. Sommerfeld, *Optics, Section 34,* New York Acacemic Press (1964)
26. D. Shindo, Z. Akase, Mater. Sci. Eng. R **142**, 100564 (2020)
27. R.K. Mishra, G. Thomas, T. Yoneyama, A. Fukuno, T. Ojima, J. Appl. Phys. **52**, 2517 (1981)
28. J.N. Chapman, A.B. Johnston, L.J. Heyderman, S. McVitie, W.A.P. Nicholson, B. Bormans, IEEE Trans. Magn. **30**, 4479 (1994)
29. S. McVitie, J.N. Chapman, L. Zhou, L.J. Heyderman, W.A.P. Nicholson, J. Magn. Magn. Mater. **148**, 232 (1995)
30. J.N. Chapman, P.E. Batson, E.M. Waddell, R.P. Ferrier, Ultramicroscopy **3**, 203 (1978)
31. K. Tsuno, Rev. Solid State Mater. Sci. **2**, 623 (1988)
32. G.R. Morrison, J.N. Chapman, A.J. Craven, *Electron Microsc.Anal.1979 Proc. Inst. Phys. Electron Microsc. Anal. Group Cof. Held Univ. Sussex Brighton 3-6 Sept.1979 EMAG 79*, CRC Press, 257 (1980)
33. K. Tsuno, M. Inoue, K. Ueno, Mater. Sci. Eng. B **3**, 403 (1989)
34. N. Shibata, Y. Kohno, A. Nakamura, S. Morishita, T. Seki, A. Kumamoto, H. Sawada, T. Matsumoto, S.D. Findlay, Y. Ikuhara, Nat. Commun. **10**, 2308 (2019)
35. M. Wu, A. Tafel, P. Hommelhoff, E. Spiecker, Appl. Phys. Lett. **114**, 013101 (2019)
36. M.R. Teague, J. Opt. Soc. Am. **73**, 1434 (1983)
37. D. Van Dyck, W. Coene, Optik **77**, 125 (1987)
38. S. Bajt, A. Barty, K.A. Nugent, M. McCartney, M. Wall, D. Paganin, Ultramicroscopy **83**, 67 (2000)
39. M. De Graef, Y. Zhu, J. Appl. Phys. **89**, 7177 (2001)
40. V.V. Volkov, Y. Zhu, Ultramicroscopy **98**, 271 (2004)
41. S. McVitie, M. Cushley, Ultramicroscopy **106**, 423 (2006)
42. D. Paganin, K.A. Nugent, Phys. Rev. Lett. **80**, 2586 (1998)
43. K. Ishizuka, B. Allman, J. Electron Microsc.Microsc. **54**, 191 (2005)
44. HREM Research Inc. http://www.hremresearch.com

Chapter 3
Application

Abstract Application of electron holography to various materials is described with classification in terms of various shapes such as particles, lines, and plates. In addition, electron holography analysis of bulk electromagnetic materials and devices is presented. Furthermore, a description is given of the electromagnetic field and collective motion of electrons obtained by in situ observation of the effects of applying various voltages and magnetic fields.

The extensive application of electron holography is presented in this chapter on the basis of the theoretical fundamentals presented in Chap. 1, and the principles and practice given in Chap. 2. Information on the electric and magnetic fields in and around various electromagnetic materials obtained by electron holography is strongly dependent on the shapes of these materials. This chapter explores the application of electron holography to various electromagnetic materials with classification in terms of the various shapes, such as particles, lines and plates. The analysis of bulk electromagnetic materials and devices using electron holography is also presented. In situ observations of electromagnetic fields and the collective motion of electrons by the application of voltages and magnetic fields are included.

3.1 Particles

One of the typical shapes employed in electron holography observations is that of particles. If the particle size is small, then direct observations of electric and magnetic fields inside the particles are possible.

Supplementary Information The online version contains supplementary material available at https://doi.org/10.1007/978-981-97-3537-2_3.

D. Shindo and Z. Akase, *Electron Holography for Electromagnetic Materials*, Springer Series in Materials Science 347, https://doi.org/10.1007/978-981-97-3537-2_3

3.1.1 Electric Field

Evaluation of the mean inner potential is presented with SiO_2 particles as one of the most basic applications of electron holography. From (2.15), when a specimen is chemically homogeneous and non-magnetic, the phase shift of the electron is related to the specimen thickness, t:

$$\phi(x, y) = \sigma\varphi(x, y)t. \tag{3.1}$$

This equation indicates that the mean inner potential can be directly determined from a measurement of the phase shift when the thickness and interaction constant are known. The specimen thickness can be determined by electron energy-loss spectroscopy (EELS) [1–4] and convergent-beam electron diffraction [5–9].

The thickness can also be evaluated from (3.1) if the mean inner potential is known. Table 3.1 lists the mean inner potentials for some typical materials [10–13]. The mean inner potential can be determined experimentally if the shape of the specimen is known. The determination of the mean inner potential for amorphous SiO_2 is demonstrated here as an example. Figure 3.1a shows an electron hologram of an amorphous SiO_2 particle with the spherical shape (Fig. 3.1b). In the enlarged hologram in Fig. 3.1c, the interference fringes shift at the particle edge. The phase shift evaluated from the interference fringes at lines X and Y is shown in Fig. 3.2. Although the phase shift in a vacuum (X) is zero, that at the particle increases significantly from the particle edge. Because the particle is spherical, the thickness of any part of the particle can be easily estimated. Dotted lines correspond to the phase shift calculated under assumed mean inner potentials of 10.5, 11.5, and 12.5 V. By comparing the experimental and calculated data, the mean inner potential for amorphous SiO_2 is evaluated to be 11.5 V. With this mean inner potential, the thickness of amorphous SiO_2 can be evaluated from holograms without the knowledge of particle shape. However, the effects of charging should be carefully considered. When relatively strong electron irradiation is used for observation, the charging effect is enhanced due to the emission of secondary electrons. The interference fringes are shifted because of the charging effect, as indicated by the arrows in Fig. 3.3a. A simulated hologram that includes the charging effect is presented in Fig. 3.3b.

The charging effect on a SiO_2 particle, with the effect of the supply of secondary electrons from the supporting film taken into account, can be studied using a mask plate as follows [14]. Figure 3.4 shows the region of interest of the specimen under electron microscope observation; a SiO_2 particle (ca. 320 nm in diameter) attached to a curled surface of the support film. Figure 3.5a shows an illustration of the geometrical configuration set for electron holography. Using a mask plate inserted in the illumination system, the illuminated area was controlled by moving its focused shadow [15]. In Fig. 3.5b, an example of a micrograph with the mask inserted is shown. The illuminated area was selected using the shadows of the mask plate and condenser aperture.

Figure 3.6a shows an electron hologram obtained when the mask was inserted. Figure 3.6b shows a reconstructed phase image obtained from the hologram of

Table 3.1 Mean inner potential of some materials in volts

C ($Z = 6$)	7.8 ± 0.6	[10]
Al (13)	13.0 ± 0.4	[10]
	12.4 ± 1	[11]
	11.9 ± 0.7	[12]
Si (14)	11.5	[13]
Cu (29)	23.5 ± 0.6	[10]
	20.1 ± 1.0	[12]
Ge (32)	15.6 ± 0.8	[12]
Au(79)	21.1 ± 2	[11]

Z = atomic number

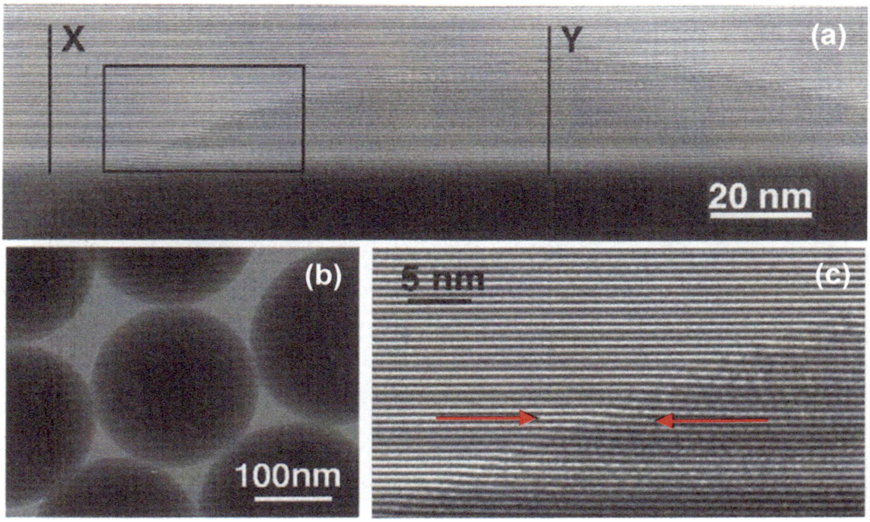

Fig. 3.1 **a** Electron hologram of amorphous SiO_2 particle. **b** Conventional TEM image of spherical SiO_2 particle. **c** Enlarged hologram of the region indicated by a rectangle in (**a**)

Fig. 3.6a, where the contour lines indicate a positively charged SiO_2 particle. Figure 3.6c–h show reconstructed phase images obtained by moving the mask shadow from left to right with movement steps of ca. 27 nm. When the support film was partially illuminated as shown in Fig. 3.6c, the contour lines around the particle did not change from those of Fig. 3.6b. When the support film was completely shielded as shown in Fig. 3.6d, the density of the contour lines around the particle increased, indicating an increase in the electric potential projected along the direction of the incident electrons. When the particle was completely shielded, charging of the particle decreased as shown in Fig. 3.6e, f. When the direction of the mask shadow movement was reversed so that the particle was again illuminated (Fig. 3.6g), it is seen that

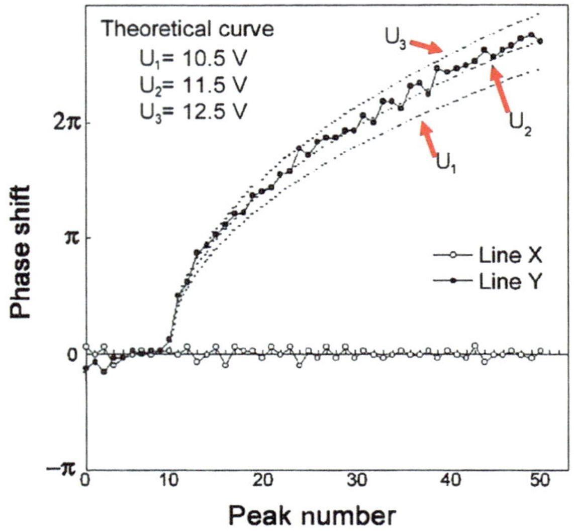

Fig. 3.2 Phase shift evaluated from interference fringes at lines X and Y

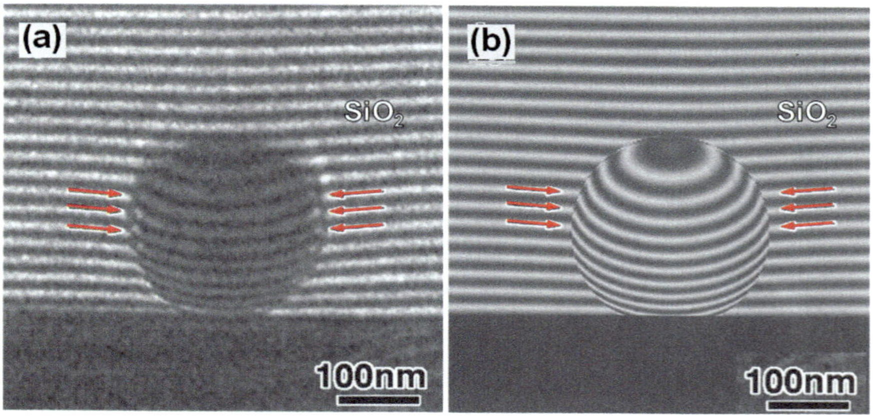

Fig. 3.3 Hologram of amorphous SiO_2 particle of 250 nm diameter (**a**) and its simulation (**b**). Arrows indicate shift of interference fringes near particle surface due to charging effect

the particle charging increased again and behavior similar to that of Fig. 3.6d was observed again. When the support film was partially illuminated again (Fig. 3.6h), the particle charging was suppressed to a level almost the same as that shown in Fig. 3.6b. These results indicate that secondary electrons emitted from the illuminated support film suppress the charging effect of the SiO_2 particle.

Simulations of the reconstructed phase images were conducted to evaluate the amount of charges on the SiO_2 particle. Figure 3.7a shows a geometrical configuration

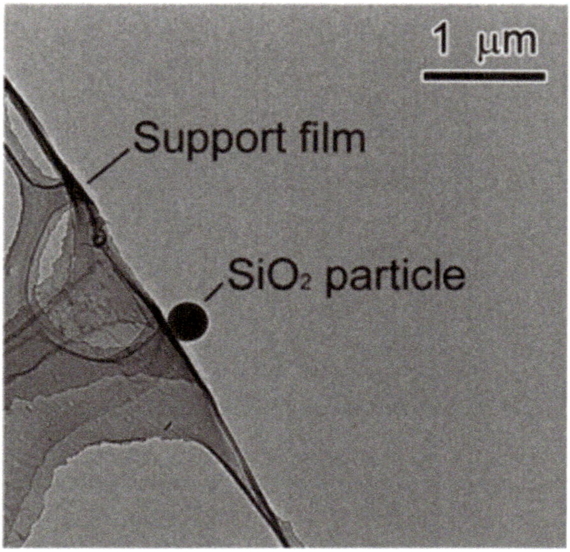

Fig. 3.4 Electron microscope image of SiO_2 particle attached to support film. Permission obtained from The Japanese Society of Microscopy [14]

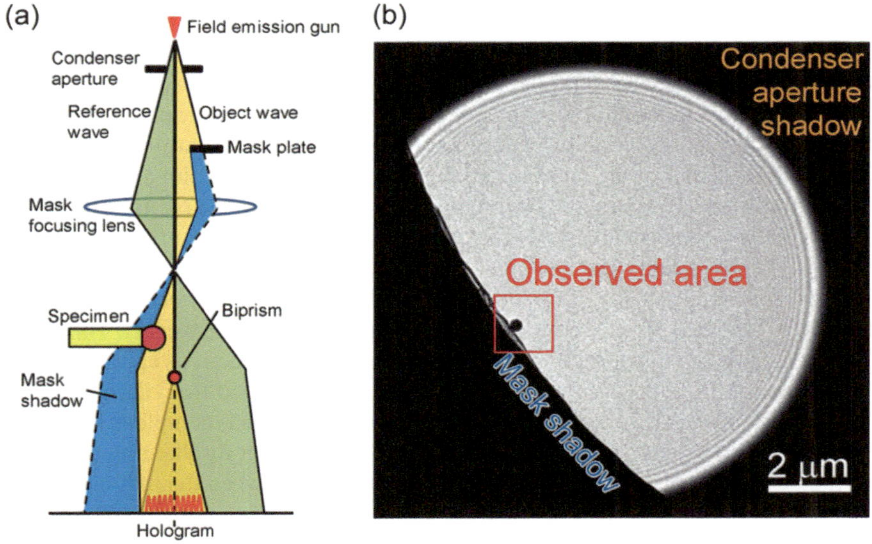

Fig. 3.5 **a** Schematic diagram of the electron optics for electron holography with mask plate. **b** Electron microscope image of electron beam illuminated area with mask shadow. Permission obtained from The Japanese Society of Microscopy [14]

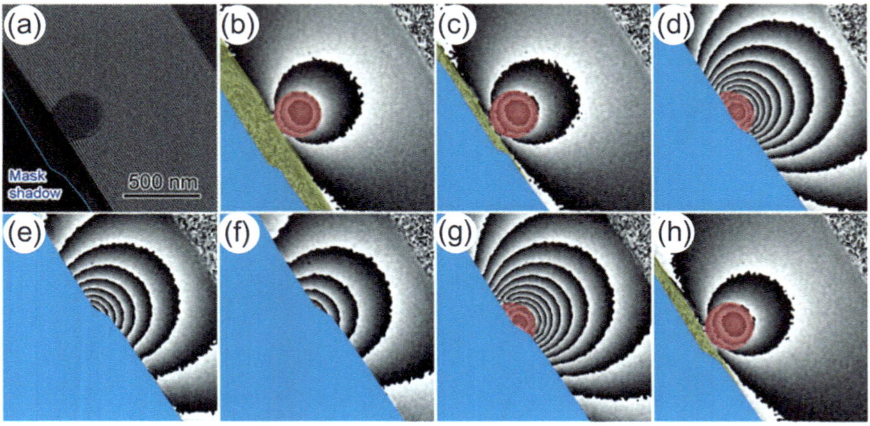

Fig. 3.6 **a** Electron hologram and **b** reconstructed phase image of (**a**). **c**–**h** Reconstructed phase images as mask shadow was moved. Permission obtained from The Japanese Society of Microscopy [14]

of the specimen model. Simulated phase images corresponding to the observation in Fig. 3.7b–h are shown in Fig. 3.7b–h where the estimated electric potentials and corresponding charge amounts are shown at the lower left corner of each figure. Figure 3.8a illustrates the experimental arrangement and Fig. 3.8b shows the charge amount as a function of the mask shadow position, where hysteresis behavior is noted. This hysteresis behavior is explained as follows. The incident electrons irradiated onto the particle induce the emission of secondary electrons, and the particle becomes positively charged accordingly. A portion of these secondary electrons emitted from the illuminated area of the particle and the conducting support film are attracted to the positively charged particle resulting in the discharge from the particle. It is noted that secondary electrons emitted from the support film significantly discharge the particle. After the mask shadow completely covers the support film, the amount of charge increases gradually. This indicates that the discharging effect caused by the secondary electrons emitted from the support film still continues weakly. The reason of this phenomenon is as follows. A portion of the incident electrons causing Fresnel diffraction hit the support film or secondary electrons emitted from the particle hit the support film causing electron emission. When the particle is completely shielded, the particle charging becomes smaller. This time dependent discharging process is attributed to a small charge flow through the interface between the particle and the conducting support film. During the backward movement, the charge amount at first increases with the increase of the irradiated area on the particle. When the position of the mask shadow reaches the area 170 nm apart from the support film, the charge amount is slowly reduced. The reason for this behavior is due to the initiation of secondary electron emission from the support film.

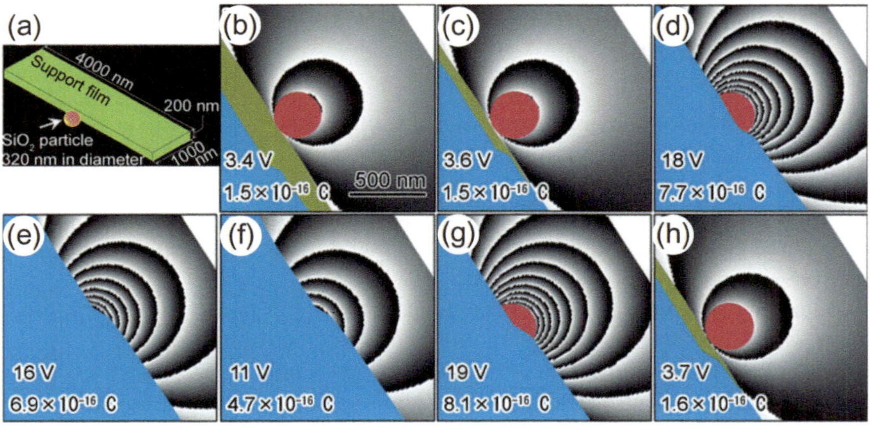

Fig. 3.7 **a** Model of the specimen used for simulation. **b–h** Simulated phase images corresponding to Fig. 3.6b–h. Electric potentials and amount of charges on the particle are shown at lower left corner. Permission obtained from The Japanese Society of Microscopy [14]

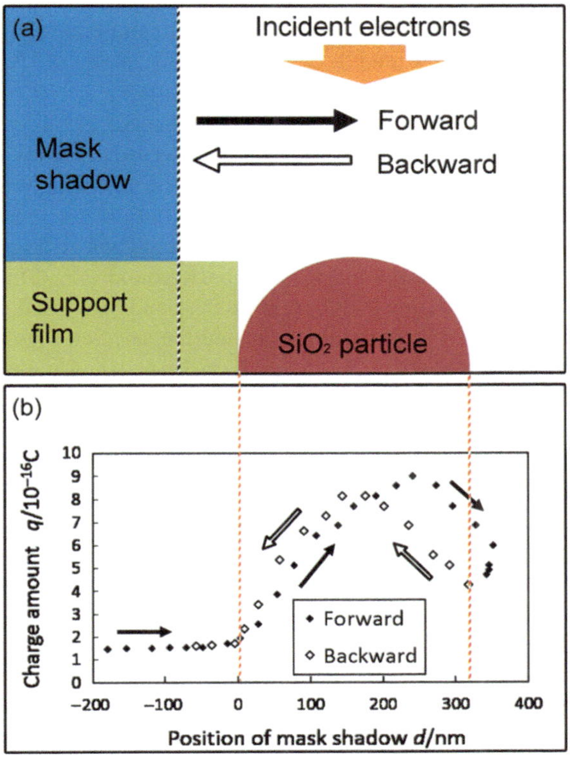

Fig. 3.8 **a** Schematic arrangement of mask movement experiment. **b** Amount of charge on SiO_2 particle as a function of mask shadow position. Hologram acquisition was conducted continuously while mask shadow was moved. Permission obtained from The Japanese Society of Microscopy [14]

3.1.2 Electric Field Fluctuation Due to Collective Motion of Electrons

When electron holograms obtained from a strongly charged specimen are reconstructed into an amplitude image, the image intensity decreases around the charged region [16, 17]. This is considered to be due to a decrease in the visibility of the hologram (interference fringes) as a result of temporal fluctuations in the electric field. These electric field fluctuations are attributed to the movement of secondary electrons around the charged specimen.

Figure 3.9 schematically illustrates the reconstruction process of an amplitude image, which includes the electron motion. The velocities of incident electrons with energies of a few hundred kiloelectronvolts are approximately 100 times higher than those of secondary electrons with energies of a few dozen electron volts. Therefore, the first incident electron is influenced by the electrostatic potential due to the charge distribution (Fig. 3.9a-1). After the first electron passes through the specimen, the following incident electrons are influenced by the electric potential due to the charge distributions ((a-2) and (a-3)). Note that the electric potential at the insulating specimen position in the lower left is stationary [16], while the electric potential along the path of the secondary electrons continually changes. Therefore, the resultant interference fringes tend to change gradually along the path of the secondary electrons, as schematically shown in Fig. 3.9b-1–b-3.

The average of the interference fringes recorded during an exposure time of approximately several seconds is observed, as shown in Fig. 3.9c-1, during which the fringe contrast along the path of the secondary electrons decreases. The digital diffractograms obtained from the path of the secondary electrons and from another area around the specimen surface are shown in Fig. 3.9c-2 and c-3, respectively. Note that the amplitudes of the background in (c-2) are large. A sideband region is selected in the reconstruction process and processed using an inverse Fourier transformation. The amplitudes outside the selected region are missing; therefore, dark contrast appears along the path of the secondary electrons, as shown in Fig. 3.9d. Note that the contrast of the reconstructed amplitude images along the path of the secondary electrons is directly related to the visibility of interference fringes (see Fig. 3.12). Color images with a color scale that indicates the relation between the visibility interference fringes and the contrast of the reconstructed amplitude images are effectively utilized, as shown in Sects. 3.2.2 and 3.3.2, to clarify the contrast change of the reconstructed amplitude images due to the electric field variations.

The time scale of the electric field fluctuations was investigated in relation to the collective motion of electrons [18]. The CMOS camera employed for in situ TEM observations (OneViewIS camera, Gatan) was utilized to continuously record holograms with short exposure times. The timescale of the electric field fluctuation was evaluated by evaluation of the intensity change of the reconstructed amplitude image due to the fluctuation of the electric field for a reduction in the exposure time.

A human kidney stone was used as an insulating specimen that would be susceptible to charging. The kidney stone was crushed and dispersed on a microgrid, which

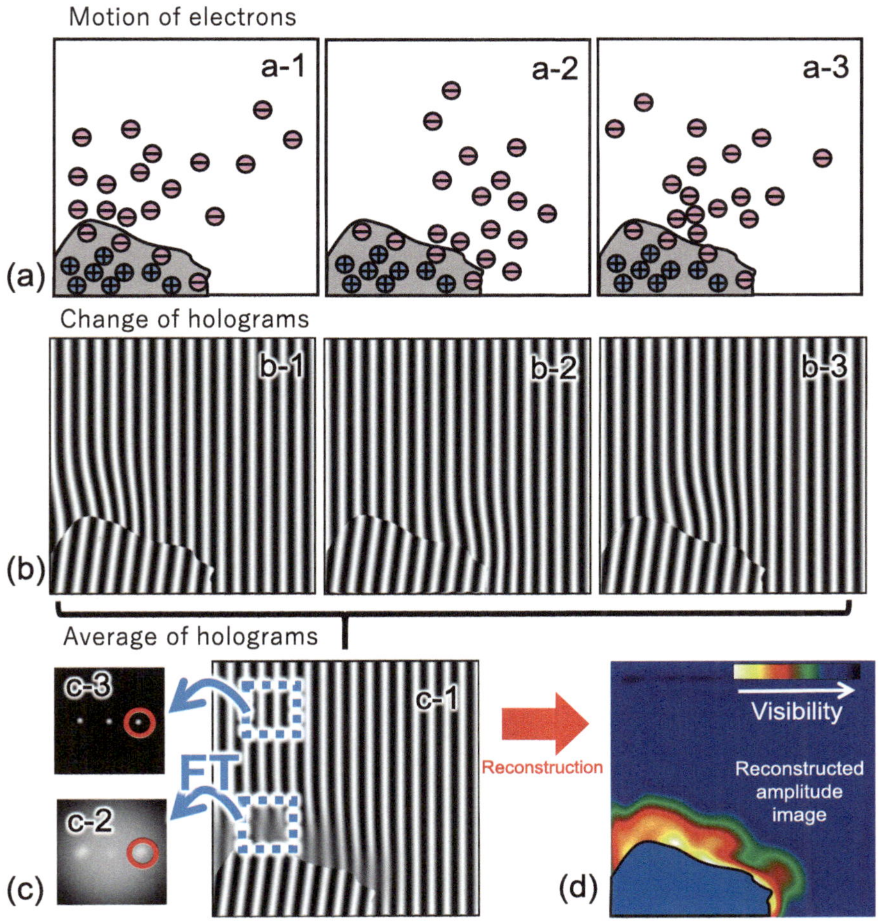

Fig. 3.9 Schematic illustration of amplitude reconstruction process for electron motions. **a** Distributions of electrons around charged specimen. **b** Change in interference fringes due to electron motions. **c** Contrast of averaged holograms (c-1). Note high amplitudes of background in digital diffractogram (c-2) obtained from region around specimen surface through Fourier transformation. **d** Amplitude image obtained from hologram (c-1). Black lines in **d** indicate specimen edge

was then cut in half. The electric field around kidney stone particles attached to the edge of the carbon support membrane was investigated. Figure 3.10a, b show a bright-field and a reconstructed phase image of the specimen, respectively. Figure 3.10b shows that the kidney stone particles were positively charged during observation. The electric potential at the specimen surface was estimated to be approximately 14 V from a simulation of the phase image.

The following procedure shown in Fig. 3.11 was used to evaluate the decrease in intensity of the reconstructed amplitude image when the exposure time was changed.

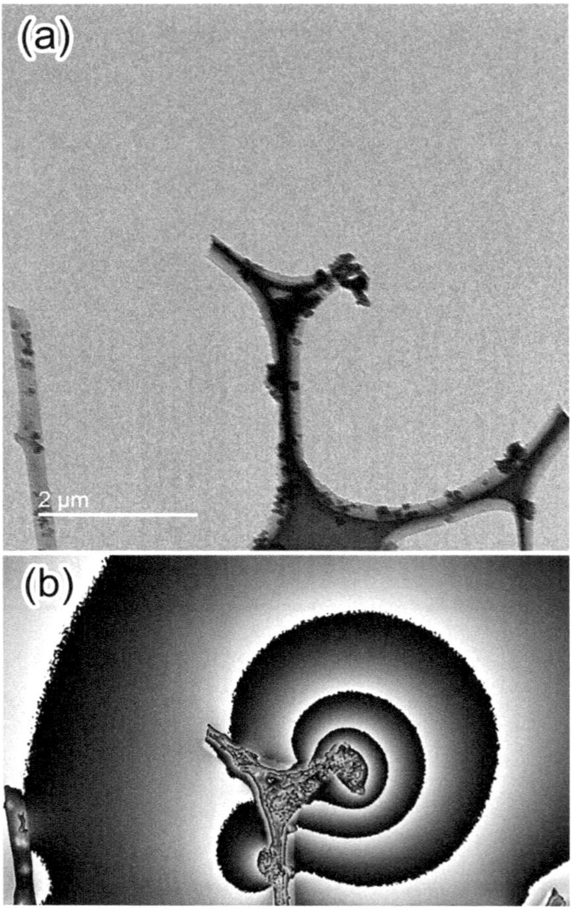

Fig. 3.10 Crushed kidney stone particles on carbon support membrane.
a Bright-field image.
b Reconstructed phase image

A series of holograms was recorded for 6 s using the in situ observation mode (4 k × 4 k pixels, 25 fps), from which three datasets of holograms, referred to as (a), (b) and (c), for amplitude reconstruction were generated by combining the frames. The exposure times for these datasets were 1.0, 0.2 and 0.08 s, and the numbers of frame-combined holograms were 6, 31 and 78, respectively. Amplitude reconstruction was performed using two different procedures, labeled (i) and (ii) in Fig. 3.11. In procedure (i), the amplitude was extracted after averaging a series of complex images, which results in an amplitude image that reflects the electric field fluctuations during the entire data acquisition time (approximately 6 s). In contrast, the amplitude is extracted from each complex image and then averaged in procedure (ii), which results in an amplitude image that reflects the electric field fluctuations during the effective exposure time.

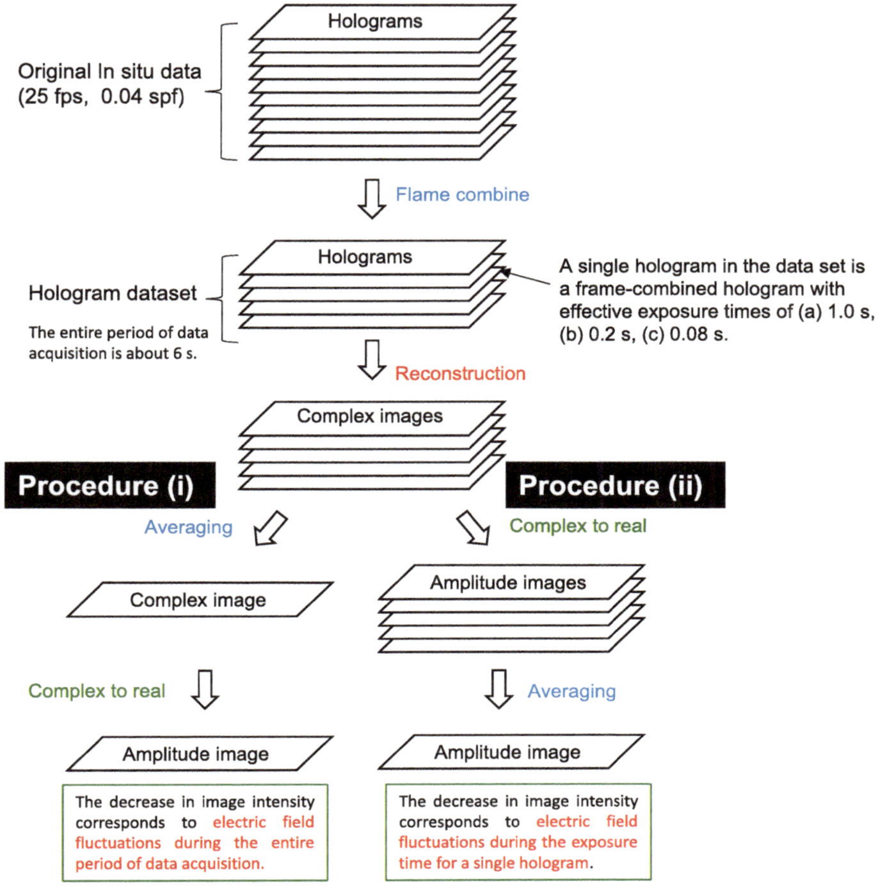

Fig. 3.11 Flow diagram for different amplitude reconstitution processes

Figure 3.12 shows reconstructed amplitude images obtained from datasets (a), (b) and (c). The upper and lower rows are for procedures (i) and (ii), respectively. As indicated by the dashed circle in Fig. 3.12a-i, there is a localized decrease in image intensity where electric field fluctuations occur. No such fluctuations are observed in Fig. 3.12c-ii, which suggests that the electric field fluctuations for this specimen mainly occurred on a timescale of approximately 0.1 s.

In this experiment, time-resolved electron holography was performed using a CMOS camera. To further improve the time resolution, it is effective to apply stroboscopic imaging methods to electron holography using a pulsed electron gun or electron beam blanking [19]. Time-resolved electron holography studies have been conducted intensively [20–22]. Application to materials characterization has been reported for $LiCoO_2$ and a Li^+-conducting solid electrolyte under cyclic voltage modulation [23].

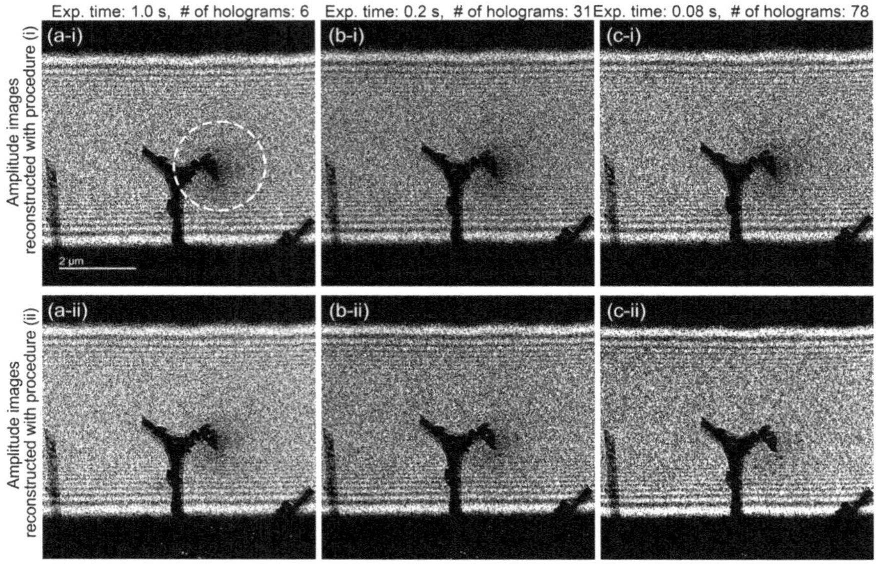

Fig. 3.12 Amplitude images reconstructed from hologram datasets (**a**), (**b**) and (**c**). Exposure time and number of holograms included in each dataset are noted at top of figure. Amplitude reconstitution procedure is different for procedures (i) and (ii) (see main text). Decrease in image intensity in upper row corresponds to electric field fluctuations during entire period of dataset acquisition. Decrease in image intensity in lower row corresponds to electric field fluctuations during exposure time for a single hologram

The time resolution in electron holography has been significantly improved using electron pulses produced by laser-driven cathodes [24]. The pump–probe technique with electron pulses have revealed a variety of dynamic behavior of materials with different TEM imaging methods; phonon vibrations in crystals have been studied using bright-field imaging [25] and the motion of magnetic skyrmions has been studied using the Lorentz method [26, 27]. Beside the laser-driven cathodes, techniques to deflect electron beams by radio frequency electric waves have been reported as gating methods for time-resolved electron microscopy [28, 29].

3.1.3 Magnetic Field

Core–Shell Co–CoO Nanocrystals

Gao et al. [30] observed chains composed of core–shell Co–CoO nanocrystals (Fig. 3.13) and reported that the magnetic thermal stability of the Co nanocrystals was significantly improved by exchange coupling between the ferromagnetic core and the antiferromagnetic shell. Although Fig. 3.13 reveals the magnetic microstructure of

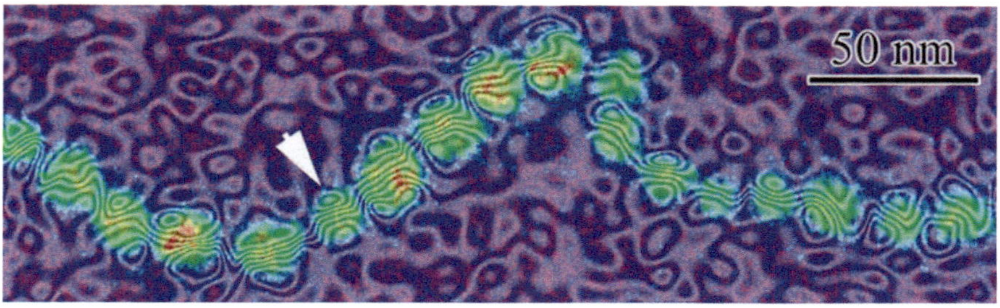

Fig. 3.13 Magnetic microstructure of a chain composed of core–shell Co–CoO nanocrystals. Permission obtained from Elsevier [17]

the chained magnetic particles, in which the lines of magnetic flux exist mainly inside the specimen (except at the end of the chain), observation of an isolated magnetic particle requires special attention. In an isolated particle that has the form of a single domain, a stray magnetic field with the direction opposite to the magnetization vector exists above and below the particle. Biskupek et al. [31] demonstrated that this stray magnetic field reduces the observable phase shift in an isolated spherical particle by a factor of $2/\pi$.

Figure 3.14 shows analysis results for the magnetic structure of Co-CoO nanoparticles. To separate information related to the electric field (mean inner potential) and magnetic field, reconstructed phase images that corresponded to two magnetic remanent states were observed by application of a magnetic field in both the left-to-right and right-to-left directions. From the difference between the two reconstructed phase images (Fig. 3.14b), the electric field indicated by ϕ_{MIP} in Fig. 3.14c was obtained. A magnetic field that corresponded to a remanent state was obtained along line M–N in (a) (Fig. 3.14d). Continuous contours were observed running through the entire particle chain, which suggests that the nanocrystal spins are aligned head-to-tail along the chain direction. This state is one possible ground state for the arrangement of magnetic nanocrystals dominated by magnetic dipolar interactions.

Fe_3O_4 Nanoparticles

Figure 3.15 shows analysis results for the magnetic structure of Fe_3O_4 nanoparticles [32]. The magnetic field information shown in Fig. 3.15a was extracted from two reconstructed phase images. One image was obtained after the specimen was flipped so that it was upside down in reference to the incident electrons. For this observation, the sense of the phase shift was reversed for the magnetic field, although it remained unchanged for the electric field. A subtraction protocol using the two reconstructed phase images erased the undesired contribution of the electric field and accordingly revealed the magnetic field (Fig. 3.15a). This figure was subjected to noise filtering, and the phase information was amplified by a factor of 100. The red lines and blue arrows indicate the outlines of individual nanoparticles and a plausible trace of the

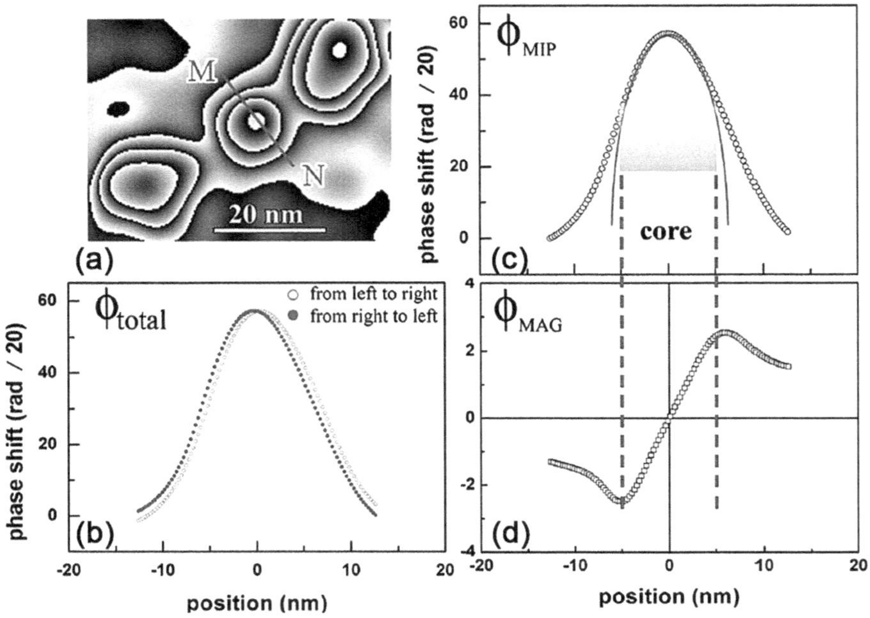

Fig. 3.14 **a** Reconstructed phase image of nanocrystal indicated in Fig. 3.13. For line M–N across the nanocrystal, plots were constructed for **b** total phase shift ϕ_{total}, **c** contribution of mean inner potential ϕ_{MIP}, and **d** contribution of magnetic induction ϕ_{MAG}. Here, ϕ_{total} was observed in two remanence states, which were saturated along the left-to-right and right-to-left directions. Diameter of Co core and thickness of CoO shell were 11.2 and 4.9 nm, respectively. Mean inner potential of core was estimated to be 18.7 V, whereas that of shell was ~16.1 V. Solid curve in ϕ_{MIP} is theoretical fit curve. Permission obtained from Elsevier [17]

easy magnetization axis, respectively. The observation clearly shows the lines of magnetic flux, i.e., the contour lines tracing the ring.

Figure 3.15b shows a simulation result for the reconstructed phase image. Figure 3.15c shows the nanoparticles were approximated by truncated octahedra with major facets of the {111} planes (indicated in green) and minor facets of the {100} planes (indicated in red). The size, location, and crystal orientation of the nanoparticles were fitted to the electron microscope observations [32]. Under the assumption that the magnetic flux density of the nanoparticles was 0.6 T [33, 34], the three-dimensional magnetic field, both inside and outside the particles, was calculated using the commercial ELF/MAGIC program [35]. The simulation reasonably explains the characteristic features of the observations. For paired particles 1–2 and 3–4, the traces of the easy magnetization axis (blue arrows) are almost parallel. For these particle pairs, the magnetic flux lines are approximately parallel to the blue arrows in both the observed (Fig. 3.15a) and simulated (Fig. 3.15b) results. In contrast, the flux lines are deformed in a zig-zag manner for particle arrays 5–6–7–8 and 10–11–1. At these locations, the blue arrows are neither mutually parallel nor

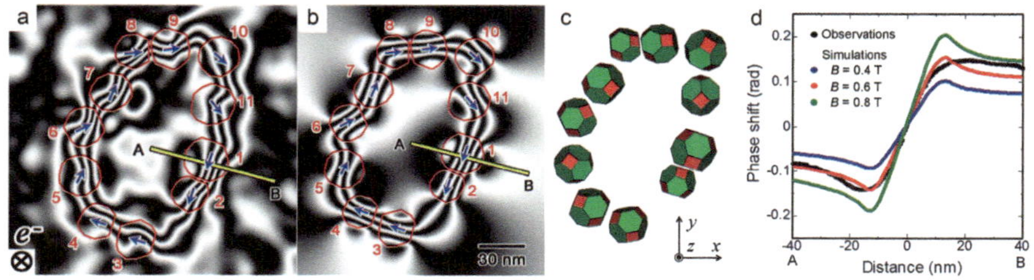

Fig. 3.15 **a** Reconstructed phase image showing magnetic information of a ring-shaped self-assembly of Fe_3O_4 nanoparticles. Phase information was amplified by a factor of 100. Blue arrows and red lines indicate traces of <111> axis and outlines of nanoparticles, respectively. **b** Simulated reconstructed phase image showing magnetic signals. **c** Structure model used in simulation. **d** Comparison between observed (black) and simulated (blue, red, green) phase shifts observed along line A–B in (**a**) and (**b**). Simulation assumed three values for magnetic flux density (B) of Fe_3O_4 nanoparticles: 0.4 T (blue), 0.6 T (red), and 0.8 T (green). Phase information was not amplified. Permission obtained from Elsevier [17]

do they trace the ring. From these observations, the fluctuations in the contour lines can be attributed to the magnetocrystalline anisotropy, which is preserved in the ring-shaped self-assembly. Note that a small discrepancy is observed in the shape of the flux lines between the observed and simulated results, as in the case of particle 9. A plausible reason for this discrepancy is an additional phase shift due to Bragg reflections [32]. The dynamical diffraction effect on the phase shift is discussed in detail in Chap. 4.

The magnetic flux density for the synthesized Fe_3O_4 nanoparticles can also be determined quantitatively. For this purpose, the phase shift was measured along the yellow line A–B that crosses particle 1 in Fig. 3.15a. These observations are plotted as black dots in Fig. 3.15d. The other curves in Fig. 3.15d show simulated results corresponding to the assumption that the magnetic flux density (B) in the particles was 0.4 T (blue), 0.6 T (red), and 0.8 T (green). These plots suggest that a flux density of 0.6 T best fits the observations. Although a small deviation (0.03 rad, between the black and red curves) is also observed, this value is comparable to the uncertainty in the phase analysis (0.03 rad), which was deduced to result from background noise. Assuming that the magnetic flux density of the nanoparticles is 0.6 T, this value corresponds to the flux density observed for bulk Fe_3O_4 crystals [33, 34]. Therefore, within the accuracy of this electron holography study, no signatures that would indicate spin disorder in the neighborhood of the surface, which would suppress the effective magnetic flux density in the nanoparticles, could be obtained.

Hollow Fe_3O_4 Spherical Submicron Particles

Electron holography was conducted here to observe the magnetic microstructure of hollow Fe_3O_4 spherical particles with a size of 420 nm. Magnetic particles with a hollow structure have arisen as an important class of nanomagnets because of the large pore volume and higher surface-to-volume ratio than the same-sized solid particles. The hollow structure results in unique magnetic features such as enhanced surface exchange bias, spin freezing, and preferential stability of the magnetic vortex.

420 nm sized hollow Fe_3O_4 spherical particles were prepared by a solvothermal method [36]. The particles are hollow spheres with averaged outer and inner diameters of 420 $\pm$ 20 and 210 $\pm$ 20 nm, respectively. Each hollow spherical particle is a mesocrystal that consists of small single-crystalline nanoparticles with almost the same crystallographic orientation. The mean crystallographic domain size of the small nanoparticles, determined from the X-ray diffraction peak profiles was 56 $\pm$ 2 nm.

Electron holograms were acquired using a 1 MV holography transmission electron microscope installed with a field-emission gun and two electron biprisms [37–39]. The phase information due to the mean inner potential and the magnetic field was separated using the so-called time-reversal operation of an electron beam [40, 41]. In this process, a pair of electron holograms were obtained for the same area, but the foil was reversed against the incident electrons. The contribution due to the mean inner potential was extracted by the addition of reconstructed phase shift images from the pair of holograms, and the phase shift due to the magnetic field was extracted by subtraction. All the observations were performed under zero applied field. The specimen was located at a magnetic-field-free position, i.e., in between the condenser lenses and the post-specimen objective lens [39], and the field strength at the position was as small as the Earth's magnetic field (<0.45G), even when the post-specimen objective lens was excited weakly to acquire electron holograms. Therefore, such a weak field is not considered to affect the magnetization of the hollow spherical particles.

Figure 3.16a, b show typical examples of unwrapped reconstructed phase images due to the mean inner potential and a magnetic field, respectively. The magnification of the images was 0.41 nm/pixel, and the voltages of the upper and lower biprisms were –165 and –110 V, respectively. The fringe spacing of the hologram was 5.9 $\pm$ 0.05 nm. The phase images due to the mean inner potential reflect the crystallographic morphology of a hollow structure, which corresponds to a bright electron microscope image [36]. On the other hand, in the case of the magnetic-field phase image, the phase shift is almost circularly symmetric and becomes larger around the particle center. Figure 3.16d shows a line profile of the phase shift due to the magnetic field contribution, which is obtained from the reconstructed phase image of Fig. 3.16b. The phase shift has a triangular-like shape with a maximum value around the particle center, which is similar to that observed for magnetic nanoparticles with a vortex structure [42]. Figure 3.16e shows the corresponding line profile of the differential phase shift, which gives information on the magnetic flux density perpendicular to the cross section given by the dotted line in Fig. 3.16b. The differential phase shift has a maximum magnitude around the position of $\pm$ 100 nm and a point symmetric shape with respect to the origin, which reflects the hollow structure and the circularly oriented magnetic flux, respectively. Such a magnetic-field phase shift is reflected by the fourfold-amplified colored magnetic induction maps shown in Fig. 3.16c. Here, the difference of two adjacent isophase lines corresponds to the twice flux

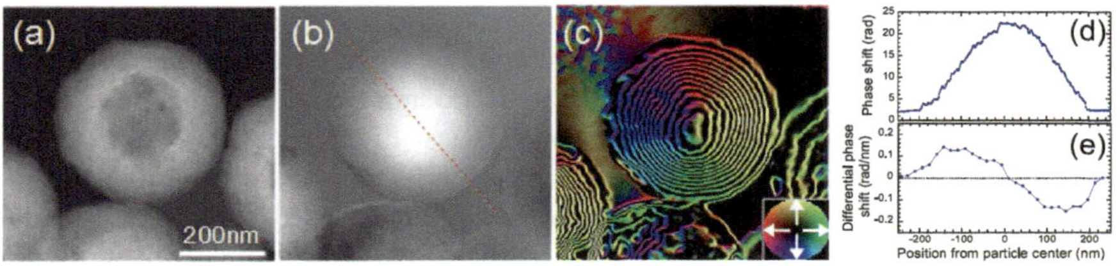

Fig. 3.16 **a** Reconstructed phase images due to the mean inner potential and **b**, **c** magnetic field, for 420 nm-sized hollow spherical particles, obtained by electron holography. **d** Line profiles of the phase shift and **e** the differential value due to the magnetic field, obtained along dotted lines in reconstructed phase image (**b**). Colored magnetic induction maps (**c**), where phase information was amplified by a factor of four. Contour lines correspond to the isophase lines, and colors shown in inset denote the direction of magnetization

quantum, (h/e). The magnetization is observed to rotate circularly within a particle to form a magnetic vortex with the core around the particle center. For electron holography imaging, only the magnetic flux perpendicular to the incident electron beam contributes to the phase shift of an electron. Therefore, magnetic information around the vortex core, which may have magnetization components parallel to the vortex axis, could not be determined in detail. The observed phase shift image due to magnetic-field contribution agrees well with the calculated phase shift image assuming a vortex structure with a vortex axis parallel to the incident electron beam. Note that the maximum value of the observed phase shift is smaller than the calculated value; this could be due to the slight deviation of spins toward the electron beam direction and/or surface spin disorder of the hollow sphere.

FeGe Tetrahedral Nanoparticles with Skyrmionic Vortex

To clarify the magnetic structure in FeGe tetrahedral nanoparticles, electron holography observation has been conducted for the [010], $[01\bar{1}]$ and $[00\bar{1}]]$ directions [43]. The reconstructed phase images represent the projections of the in-plane magnetic fluxes integrated along the beam trajectory. Figure 3.17a shows an SEM image of a 145 nm FeGe tetrahedral nanoparticle on a probe. Electron holography was performed at 10 K. Electron microscope observations along the three directions are shown in Fig. 3.17c–e. Systematic electron holographic imaging reveals a swirling in-plane magnetic flux along the $[00\bar{1}]$ direction (Fig. 3.17f), a distorted swirling in-plane magnetic flux along the $[01\bar{1}]$ direction (Fig. 3.17g), and an unconventional magnetic flux distribution along the [010] direction (Fig. 3.17h). Figure 3.17i–k show the corresponding projections for the magnetic phase distribution calculated from micromagnetic simulations in a tetrahedral particle of the same size. The simulation captures the essential features of the observed magnetic texture well. The difference between the phase shift profiles along the lines in Fig. 3.17f and i is within reason (Fig. 3.17b). The phase shift in the simulation is underestimated because it does not account for the additional magnetic phase shift from the amorphous surface layer. The simulated 3D magnetic configuration, herein referred to as the skyrmionic vortex, is detailed in Fig. 3.17l–n. The non-triviality is indicated by the tube-like equi-spin surface. The equi-spin surface with the z component (along the [001] direction) of magnetization $Sz = 0$ forms a tube that stretches along the [001] direction (Fig. 3.17m). The in-plane magnetic configurations also show that the tube in the skyrmionic vortex state is formed in each cross section (Fig. 3.17 m and n). The projected magnetic configuration along the [00 $\overline{1}$] direction (Fig. 3.17l) resembles a vortex state with unity vorticity, in which the in-plane magnetic moments rotate by an angle of 2π circulating around the centre. A unity vorticity state can be either a singularity, meron or skyrmion. The magnetic moments here are well defined everywhere; therefore, they cannot be a singularity.

Furthermore, unlike meron states, the magnetic moments herein flip their z components from the centre to the periphery. The emergence of skyrmionic texture in this tetrahedron is further evidenced by the topological charge calculation. Confined geometries generally have finite boundaries and the topological charge is no longer strictly integer-valued. By excluding the corner spin twists, the total topological charge enclosed by the dashed box in Fig. 3.17l has a fractional value of 0.93. The slight deviation from 1 is mainly attributed to entanglement with corner spin twists at the boundary of the skyrmionic vortex.

The skyrmionic vortex tube exhibits a richer structure than conventional skyrmion tubes. Although the core spins of the skyrmionic vortex pass uniformly through the

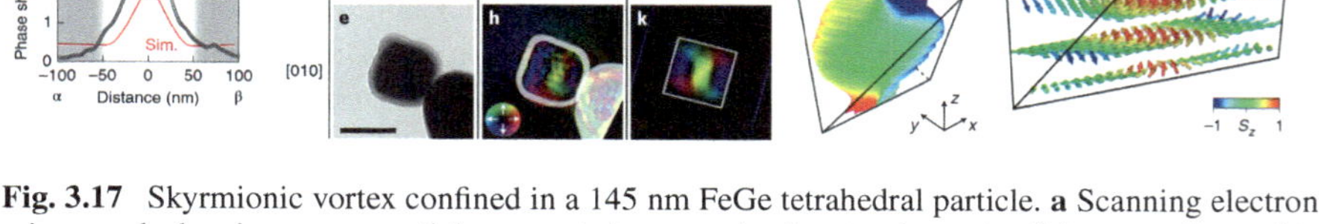

Fig. 3.17 Skyrmionic vortex confined in a 145 nm FeGe tetrahedral particle. **a** Scanning electron micrograph showing a nanoparticle supported on a probe. Inset: schematic of the crystallographic orientation. **b** Phase-shift profiles along the lines in f and i. Sim: simulated. **c–e**, TEM images along [00$\overline{1}$] (**c**), [01$\overline{1}$] (**d**) and [010] (**e**) directions. **f–h**, Observed in-plane magnetic flux distributions for corresponding projections at 10 K. Masked regions cover fringes and probe, where phase reconstruction or subtraction fails because of abrupt change in traversing thickness, poor visibility and/or slight deviation in projection conditions at different temperatures. Colour wheel in h represents direction and magnitude of projected in-plane magnetic flux. i–k, Corresponding projections of micromagnetic simulations. l, Magnetic configuration of i integrated along [00$\overline{1}$]direction. m, Equi-spin surface with z component (z = [001]) of magnetization Sz = 0. n, Magnetic configurations in cross-sectional planes normal to z-axis. Scale bars, 100 nm. Permission obtained from Nature Research [43]

[001] direction, the peripheral swirling pitch does not. A Bloch-like rotation with a wider pitch is present in the central cross section, a Neel-like rotation with a narrower pitch is emergent near the top and bottom, and mixed helicities are formed in between, which altogether give rise to the appearance of a potbelly-shaped equi-spin surface. In addition, a topological defect known as a fractional vortex also forms on the edge due to chiral geometric frustration, i.e., the interplay between the geometrical confinement of the tetrahedron and the bulk helimagnetic interaction. The two fractional vortices on the top and bottom edges are connected through a vertical line with spins that point in the [001] direction (Fig. 3.17n) and facilitate the formation of the skyrmionic vortex tube.

With this portrait of the skyrmionic vortex, the observed and simulated projections (Fig. 3.17f–k) can now be readily interpreted as follows: the [00 $\bar{1}$] projections (Fig. 3.17f, i) are identical to the top view of this skyrmionic vortex and capture the swirling contrasts. The [010] projections (Fig. 3.17h, k) are the side view, with the [001] polarity of the core color-coded with yellow bars. The [01$\bar{1}$] projections (Fig. 3.17g, j) are intermediate projections that mix the features of the top and side views and thus reveal a distorted swirling texture. It is probable that a conventional magnetic vortex structure could yield similar projections. However, according to the comparative simulations for a vortex structure, the [010] projection of a vortex shows a significant contrast in a darker and narrower spine component compared with that of a skyrmionic vortex, which enables a skyrmionic vortex to be distinguished from a pure vortex. Therefore, the possibility of a pure vortex state has been ruled out.

3.2 Lines

3.2.1 Electric Field

Field Emission from $TaSi_2$ Nanowire

In situ studies under field-emission conditions have been extensively conducted on the electron emitters utilizing nanotubes and nanowires to determine their properties [44–51]. Here we present the experimental setup and the obtained *I–V* curve used to study the field emission from a single $TaSi_2$ nanowire [52] in Fig. 3.18a. The *I–V* curve was recorded using a transmission electron microscope. When the voltage was increased to approximately 50 V, a certain emission current (*ca.* 1 nA) was detected. The reconstructed phase images in Fig. 3.18b show the change in the electric potential near the $TaSi_2$ nanowire with an increase in the applied voltage. The observations represent the electric field projected along the incident electron beam. The spacing between the equipotential lines decreases as the applied voltage is increased. The reconstructed phase image is highly blurred at a voltage of 50 V. In Fig. 3.18b, the equipotential lines shown are approximately parallel to the anode edge, and the peculiar modulation of contour lines is difficult to observe because of the charged emitters as reported by Matteucci et al. [53] and Cumings et al. [51]. This phenomenon is observed because a large electric field was applied between the anode and cathode, which were both substantially larger than the nanowire in this experiment. It is also noted that the phase shift due to the tiny nanowire should be much smaller than that due to the anode and cathode.

Figure 3.19a, b show the reconstructed phase image obtained at an applied voltage of 50 V and a simulated image, respectively. In Fig. 3.19a, a region of irregular contrast is observed between the W anode and the $TaSi_2$ nanowire, as outlined in green. The simulated image was obtained by assuming a bias voltage of 50 V with a fluctuation of 1 V. From a comparison of the observed and simulated images, it was

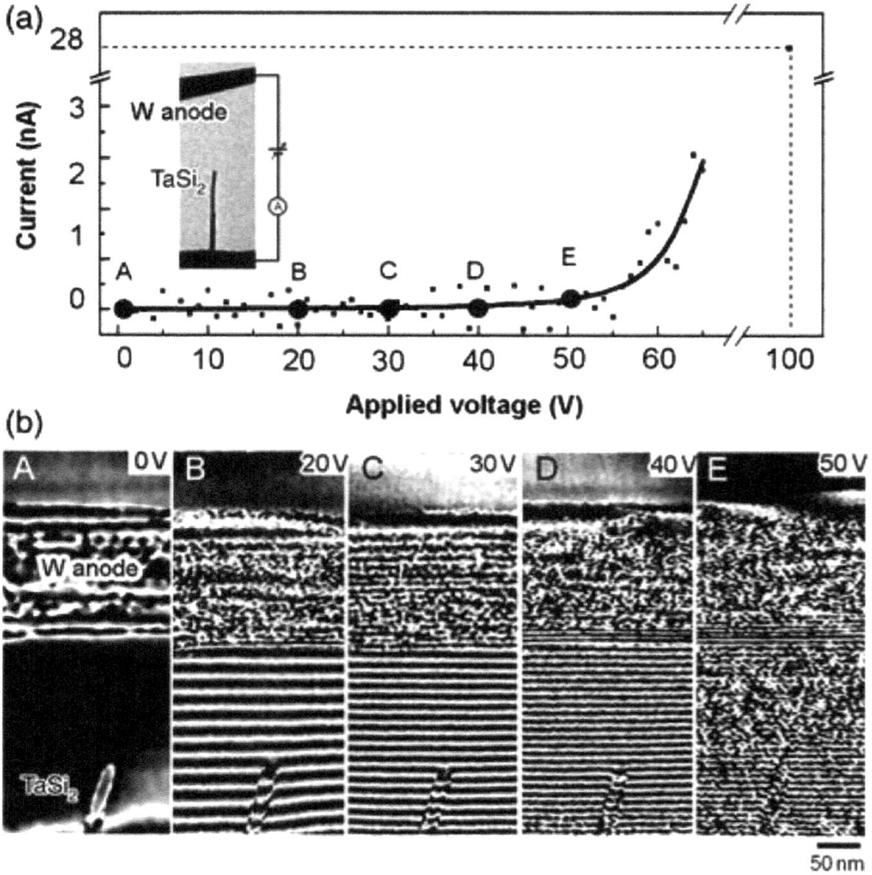

Fig. 3.18 **a** I–V curve showing field emission from a $TaSi_2$ nanowire, observed using experimental setup shown in inset, and **b** changes in reconstructed phase images with increasing applied voltage. Permission obtained from Elsevier [17]

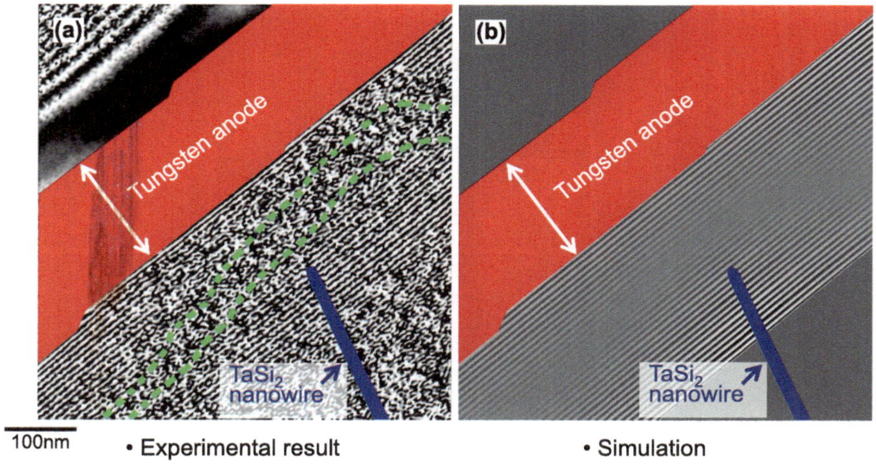

Fig. 3.19 **a** Reconstructed phase image obtained from a $TaSi_2$ nanowire at applied voltage of 50 V and **b** simulation result for comparison. Permission obtained from Elsevier [17]

concluded that the irregularity in contrast resulted from the fluctuation of the electric potential in the nanowire due to ballistic emission [52].

Although the detailed electric field around the field emitter was clarified by application of a bias voltage, the electric field variation due to the motion of electrons emitted from the field emitters could not be detected, because the electric field variation due to the emission current of *ca.* 0.2 nA was too small to detect.

Electric Field Around Schottky Field Emitter

Evaluation of the electric potential around field emitters is important to understand the emission properties. In particular, for electron holography studies, an electron beam with high coherence can be produced from a field-emission gun (FEG). Matteucci et al. [53] performed an incipient study of the potential distribution near a tungsten microtip. Cumings et al. [51] showed a reconstructed phase image that represented the field emission from a carbon nanotube. Herein, observations on the electric field around Schottky emitters are presented.

Figure 3.20 shows SEM images of the unused and used emitters. A Schottky FEG-emitter is also referred to as a W–ZrO emitter; the emitter body is made of tungsten (W) and this is coated with zirconium oxide (ZrO_2) to reduce the work function of the emitter. Figure 3.20a, b show a lump of ZrO_2 on the body of a used emitter, as well as that of an unused emitter. The lifetime of the used emitter is exhausted by operation of approximately 10 000 h. Therefore, it is assumed that the cessation of emission in this emitter did not result from complete exhaustion of the ZrO_2 lump.

Enlarged images of the top edges of the two emitters are shown in Fig. 3.20c, d. In the case of the used emitter, no severe mechanical modification was observed, such

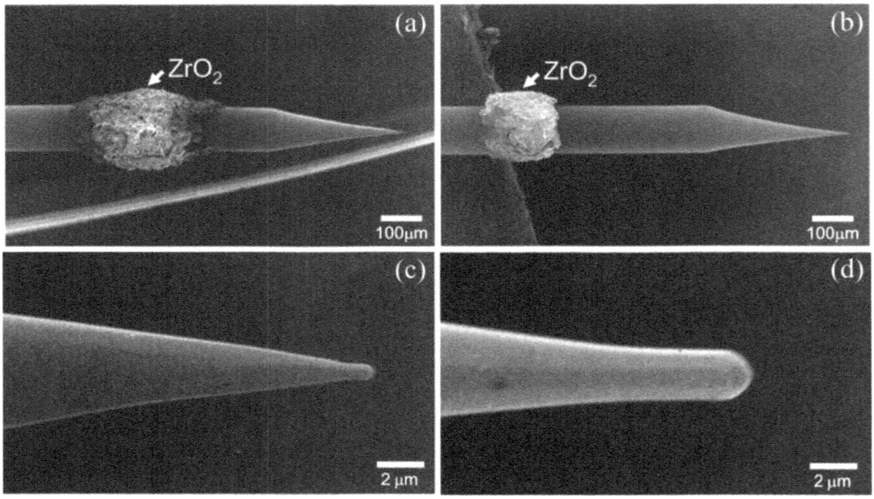

Fig. 3.20 Scanning electron microscope images showing unused (**a**, **c**) and used (**b**, **d**) emitters. Permission obtained from the Japanese society of microscopy [54]

as shape deformation and whisker growth. However, it a clear change in the radius of curvature occurred due to the repeated emission; the measured radius of curvature of the used emitter is *ca.* 1.2 μm, while that of the unused emitter is *ca.* 0.4 μm.

Figure 3.21a–d show electron holograms (upper panels) and reconstructed images (lower panels) obtained from unused (a, c) and used (b, d) Schottky emitters at an applied voltage of 100 V [54]. A difference in the edge shapes is evident—specifically, a deformation in the W(100) facet of the used emitter. This observation implies that sputtering effects of the residual ions under the influence of the large electric field caused the W(100) facet to wear. Accordingly, the electric potential differs around the field emitters, as shown in Fig. 3.21e, f, which show enlarged images of the areas marked in Fig. 3.21c, d, respectively. The difference in the electric potential distribution is clear in Fig. 3.21g, which shows plots of the phase shift along each *X*–*Y* line. Variations in the electric potential distribution around a cold-type FEG emitter were also investigated using a similar analysis method [55].

3.2.2 *Visualization of the Collective Motion of Electrons Around Microfibrils*

Visualization of Electron Orbits

Figure 3.22 shows electron microscope images of a microfibril of sciatic nerve tissue observed under over-focus conditions. Around a line-shaped insulating specimen,

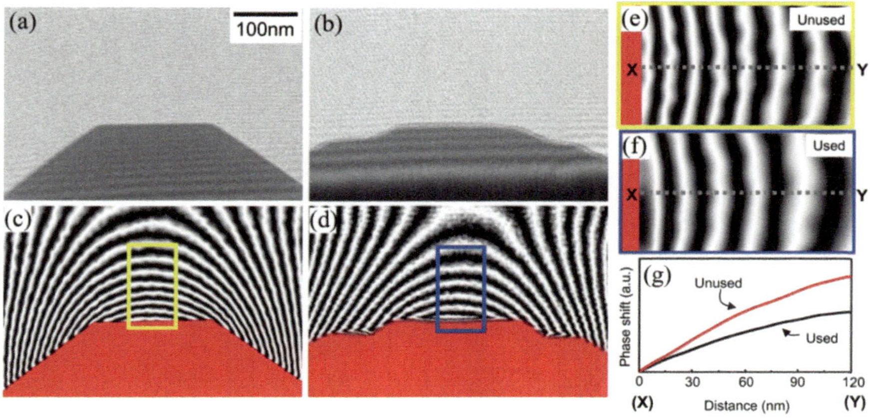

Fig. 3.21 Electron holography results showing differences in electric potential distributions around unused and used emitters. (**a**, **c**) and (**b**, **d**) are electron holograms and corresponding reconstructed phase images in case of unused and used emitters, respectively. Emitters are denoted by color red. In (**c**) and (**d**), regions indicated by colors yellow and blue are shown in (**e**) and (**f**), respectively. Phase-shift variations were measured along grey broken lines (X–Y) indicated in (**e**) and (**f**), and this result is shown in (**g**). Length in grey broken lines (X–Y) in (**e**) and (**f**) is the same with that of x-axis in the graph of (**g**). Permission obtained from the Japnese society of microscopy [54]

the incident electron beam was deflected by the positively charged specimen. The contrast in the top-edge region changed significantly as the intensity of the beam was increased. As indicated by the inset illustrations, this contrast change is interpreted as enhanced deflection of the incident electron beam due to the increased charging effect [56]. The stationary orbits of secondary electrons result from this enhancement of the charging effects in the top-edge region in line-shaped specimens, as described below.

Figure 3.23a, c, d respectively show a hologram, a reconstructed phase image, and a reconstructed amplitude image of three entangled microfibrils, respectively [57]. The intensity of the incident electron beam for observation of the hologram was 60 $nm^{-2} \cdot s^{-1}$. The electric potential at the specimen surface was evaluated to be 30 V by computer simulation [58]. There are several irregular contrasts in the reconstructed phase image (Fig. 3.23c) that result from variation in the electric field due to the motion of secondary electrons. It is also noted that the hologram has interference fringes missing in regions A, B, and D (Fig. 3.23b), which results in large amplitude of the background in the digital diffractograms, whereas there are no missing fringes in region C.

The reconstructed amplitude image of Fig. 3.23d shows several circuits with dark contrasts which correspond to the orbits of secondary electrons around these regions. In the enlarged hologram of region E, the extrapolated dotted lines along the interference fringes shift away from each other in the regions without interference

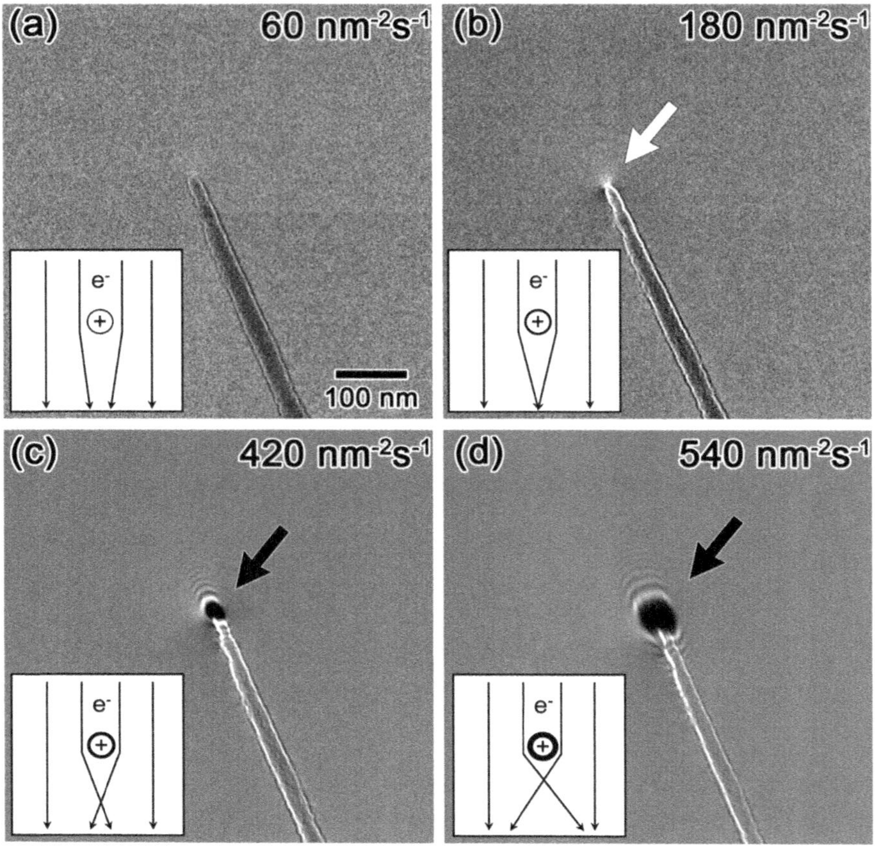

Fig. 3.22 Electron microscope images of microfibril of sciatic nerve tissue observed under over-focus condition. As intensity of incident electron beam was increased, contrast in top-edge region changed drastically. Inserted illustrations show deflection of incident electron beam due to charging effect in top-edge region. Permission obtained from Elsevier [17]

fringes [57]. It was reported that the interference fringes in the hologram disappeared completely when the projected density of fluctuating secondary electrons was approximately 1 nm^{-2} for incident 200 and 300 keV electrons [57, 59].

Observation with a single microfibril was carried out for comparison with the stationary electron orbits around the tangled microfibrils. The bright-field image in Fig. 3.24a shows the surface of the microfibril was uneven with some dark spots indicated by blue arrowheads. The dark spots are considered to be Os-rich regions.

The uneven surface and inhomogeneous distribution of the coating element would likely cause an inhomogeneous charge distribution on the surface of the microfibrils of sciatic nerve tissue resulting in the characteristic motion of collective secondary electrons. The reconstructed phase image in Fig. 3.24b, which was obtained at an

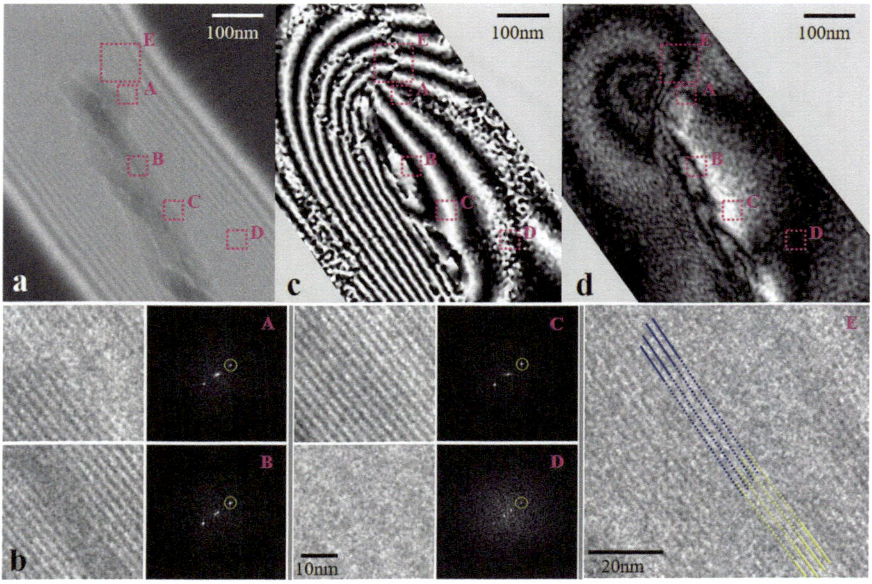

Fig. 3.23 **a** Electron hologram of tangled microfibrils of sciatic nerve tissue. **b** Enlarged images of rectangular regions (A–E) indicated by broken pink lines in **a** together with digital diffractograms (A–D). Aperture size for reconstruction of amplitude image is indicated by yellow circle in each digital diffractogram. **c** Reconstructed phase image. **d** Reconstructed amplitude image. Rectangular regions (A–E) are also presented in (**c**) and (**d**) for comparison. Permission obtained from Journal of the Physical Society of Japan [57]

electron intensity of 30 $nm^{-2} \cdot s^{-1}$, shows an electric field around the microfibril. In the reconstructed amplitude image in Fig. 3.24c, there is a slightly dark contrast on the right side of the microfibril indicated by a short yellow arrow. When the incident electron intensity was increased to 36 $nm^{-2} \cdot s^{-1}$, the dark-contrast region shifted to the left of the microfibril in Fig. 3.24d.

As indicated by the narrow red arrow, a sharp dark-contrast region was observed along the microfibril. In a few tens of seconds, the dark-contrast region formed a circular shape around the top of the microfibril as shown in Fig. 3.24e. The sharp line contrast was still visible, as indicated by the narrow red arrow. After another few tens of seconds, the circular shape became larger and shifted to the right as shown in Fig. 3.24f. The change in its shape around a single microfibril indicates that the rather unstable motion of collective secondary electrons differs from that observed with tangled microfibrils.

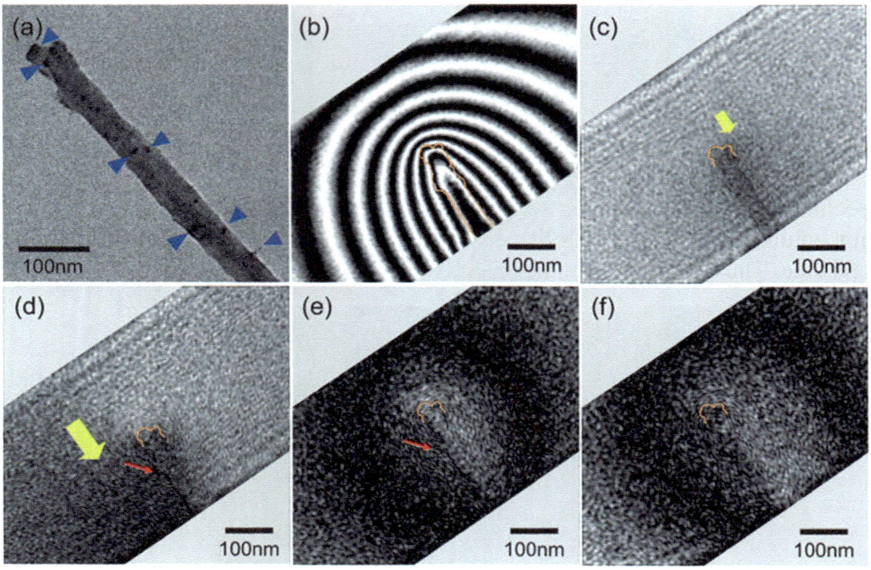

Fig. 3.24 **a** Bright-field image of single microfibril of sciatic nerve tissue. **b, c** Reconstructed phase and amplitude images, respectively, of microfibril at incident electron intensity of 30 $nm^{-2}{\bullet}s^{-1}$. Corresponding parts of microfibril are outlined in brown. **d–f** Reconstructed amplitude images at incident electron intensity of 36 $nm^{-2}{\bullet}s^{-1}$. Permission obtained from Journal of the Physical Society of Japan [57]

Change in Electron Orbits Due to Insertion of Electrode

In Fig. 3.25a, when a small electrode is inserted near the microfibrils, the small electric circuit around the top region of the microfibrils and the circuits around region D in Fig. 3.23d disappear, leaving a few simple circuits due to the change in the electric field. When a bias voltage of 5 V is applied to the electrode, the circuit becomes smaller in Fig. 3.25b. Although the sizes and shapes of the electric circuits change due to the field produced by the inserted electrode, one of the regions of secondary electron spread is located at the same position, as indicated by the red arrowheads.

Visualization of Accumulative and Collective Motion of Electrons

In Fig. 3.26a, an electron microscope image shows a microfibril with uneven surfaces ending in two branches. The hologram of Fig. 3.26b corresponds to the top right region indicated by the white square in Fig. 3.26a [60]. Note that the incident electron intensity was around 0.15 $nm^{-2} \cdot s^{-1}$, which is much smaller than the case of electron orbits discussed above.

A reconstructed amplitude image at the initial state is presented in Fig. 3.27a. Figure 3.27b, c show reconstructed amplitude images with the passage of time [60].

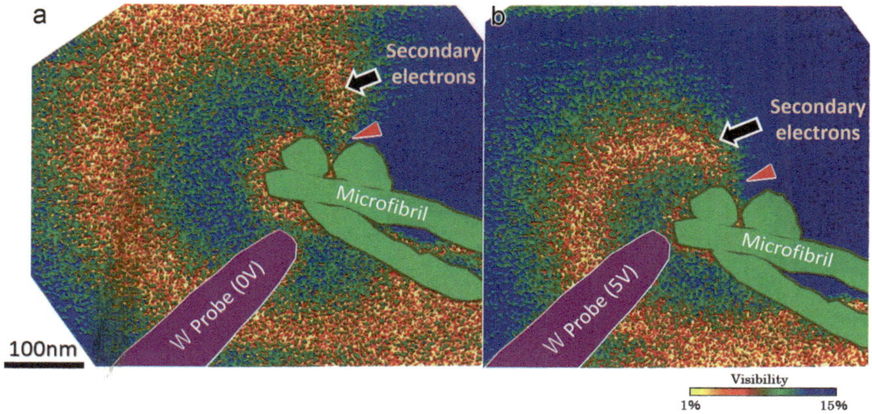

Fig. 3.25 Reconstructed amplitude images showing change in electric circuits of secondary electrons resulting from insertion of electrode. Color bar indicates visibility of fringe contrast in original hologram. **a** Electrode (tungsten) outlined in white; bias voltage of electrode is zero. **b** Bias voltage of electrode is +5 V. Permission obtained from Elsevier [17]

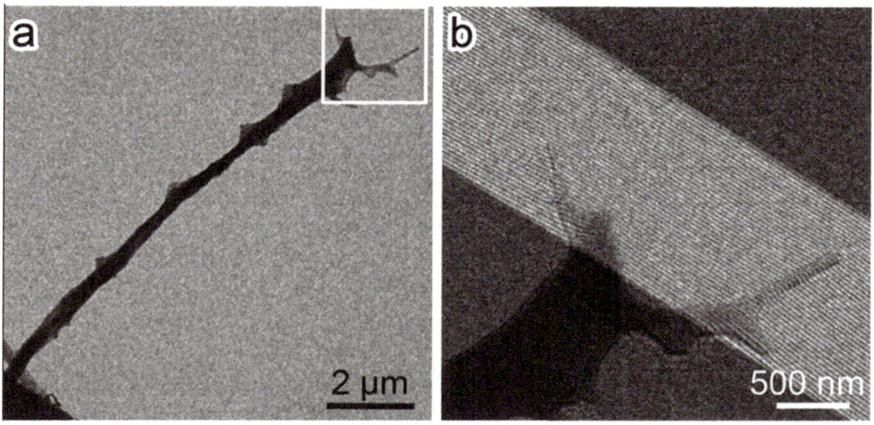

Fig. 3.26 **a** Electron micrograph and **b** hologram of microfibril of sciatic nerve tissue observed at initial state. Permission obtained from Elsevier [17]

The bright red regions, which are considered to correspond to the fluctuation of the electric potential due to the motion of the secondary electrons, gradually shift between the branches. Contrast formation in the reconstructed amplitude images due to the accumulation of electrons around charged microfibrils was also observed in a similar manner to that in Fig. 3.9 [59]. Figures 3.28 and 3.29 show video captures of the collective motion of electrons in black and white and in color, respectively.

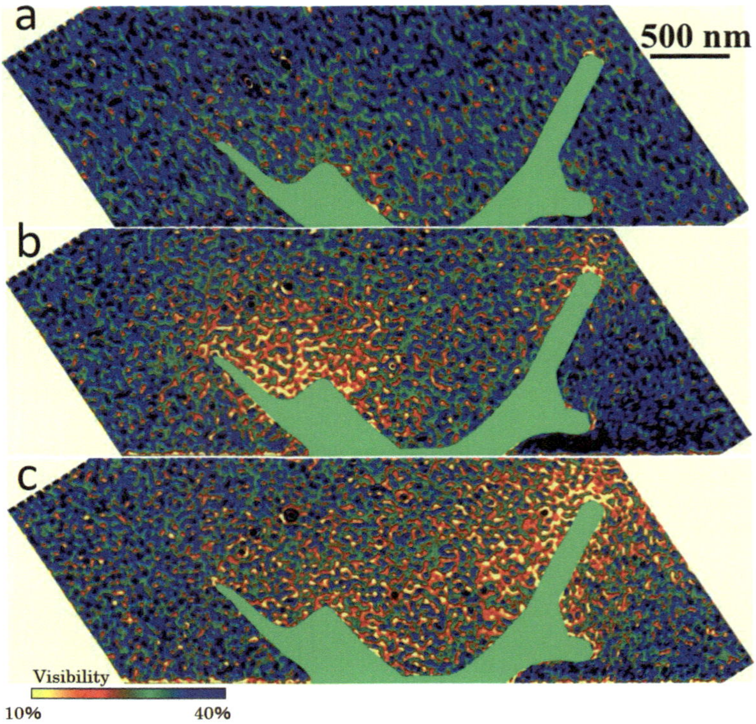

Fig. 3.27 **a** Reconstructed amplitude images obtained from hologram of microfibril at initial state shown in Fig. 3.26b. **b**, **c** Amplitude images reconstructed from holograms recorded with passage of time from initial state. Permission obtained from Elsevier [17]

3.2.3 Magnetic Field

The magnetic interaction between the pole tip of a single-pole head and a pseudo soft underlayer (SUL) in a perpendicular magnetic recording head is presented [61]. Figure 3.30 shows a bright-field image of the specimen. The thin Ru layer can be observed as a dark contrast at the center of the amorphous-like SUL; the other parts appear in strong contrast, which indicates their crystalline microstructure. Figure 3.31a shows a reconstructed phase image with the lines of magnetic flux going outward from the pole tip, which is excited by the applied field and absorbed into the SUL. Some areas show slight discontinuities in the lines, as indicated by red circles, which are considered to be due to the significant change in the inner potential at the boundaries. In particular, the lines of magnetic flux inside the SUL have a zigzag appearance, which is highlighted in the region enclosed by a pink circle; this caused some difficulty in interpretation. To remove these influences of inner potential, another hologram was taken after turning the specimen over with respect to the incident electron beam [40]. The mean inner potential was removed by

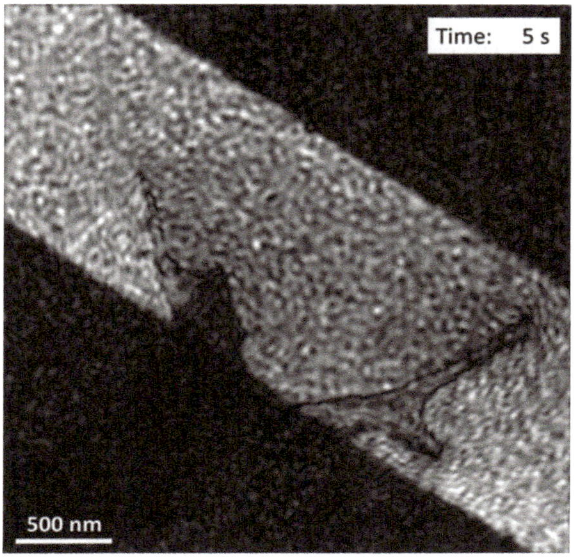

Fig. 3.28 Video: Collective motions of electrons in black and white

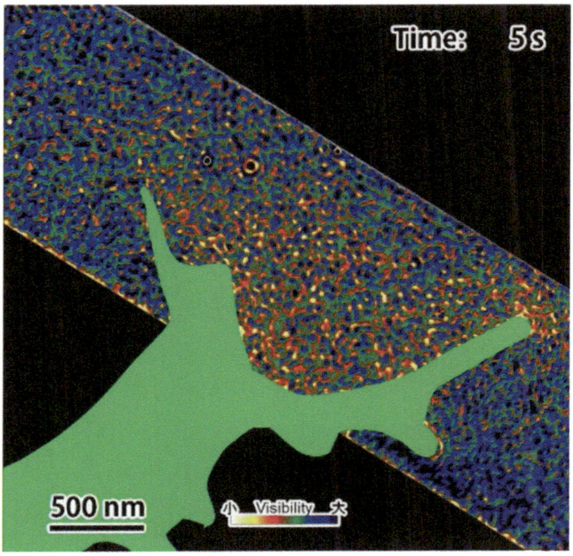

Fig. 3.29 Video Collective motions of electrons in color

subtracting the phase information for one hologram from the other, which left only the magnetic information. Figure 3.31b shows that the discontinuities have disappeared and the lines of magnetic flux are more clearly observed. The lines inside the SUL became curves in one case only, which indicates the synthetic antiferromagnetic coupling interaction between the upper and lower parts of the SUL. Non-negligible stray fields exist in the vacuum region in Fig. 3.31b, despite the presence of the NiFe shield layer. These stray fields are considered to have an effect on the reconstructed phase image and the quantitative evaluation of magnetic flux density inside the SUL.

Figure 3.32 shows a bird's-eye view of the mesh partitioning of the finite element model (FEM) on the same scale as the specimen, in which the Al_2O_3, carbon, and Ti layers are dealt with as being equivalent to air, with permeability μ_0. The saturation magnetizations of the main pole, SUL, and NiFe are set to be the measured values of 2.0, 1.3, and 0.9 T, respectively. The Z-direction is the direction of the incident electron beam.

Figure 3.33a shows the calculated reconstructed phase image. Note that this image would correspond to the experimental reconstructed phase image when the modulation of the reference wave is exactly zero, with the supposition that such an experiment could be carried out. With reference back to the experimental image in Fig. 3.31, the positions of the biprism and the regions of the object wave and the reference wave were determined. Figure 3.33b shows the reconstructed phase image obtained by considering the modulation of phase shift in the reference wave. Figure 3.33a shows that the magnetic flux densities in all materials (the pole, SUL, NiFe) and the stray fields are larger than those in Fig. 3.33b, which implies that the magnetic flux densities in Fig. 3.31b obtained experimentally were underestimated with respect to their true values. The directions of the magnetic flux around the pole tip in Fig. 3.33b are observed to change to some extent in comparison with Fig. 3.33a; this was due to the relatively strong magnetic field in the reference region.

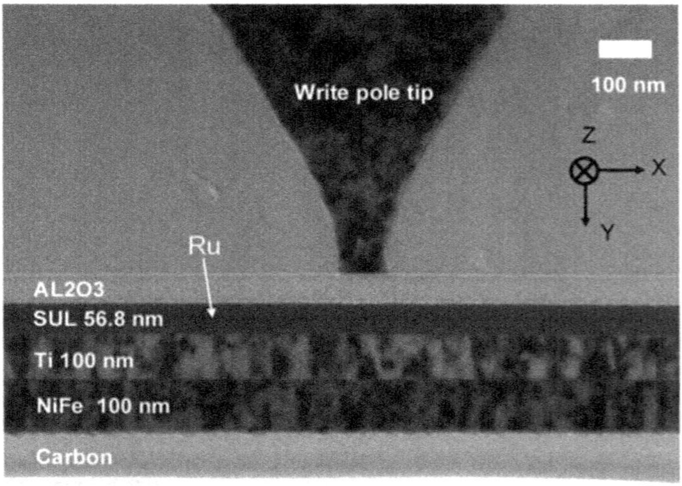

Fig. 3.30 Cross-sectional bright field image of the pole tip and pseudo SUL. Thick NiFe layer was designed to be a magnetic shield layer. Permission obtained from The Japanese Society of Microscopy [61]

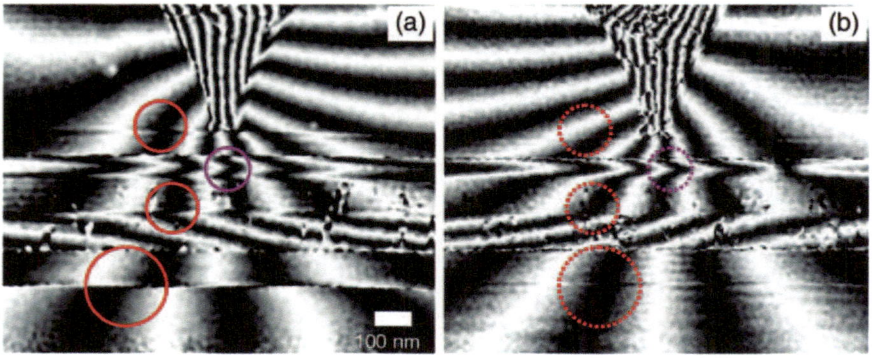

Fig. 3.31 **a** Reconstructed phase image including both magnetic and electrical (inner potential) information. Circles indicate the disconnections of lines of magnetic flux due to inner potential. In (**b**), inner potential has been removed, and dashed circles correspond to solid circles in (**a**). Permission obtained from The Japanese Society of Microscopy [61]

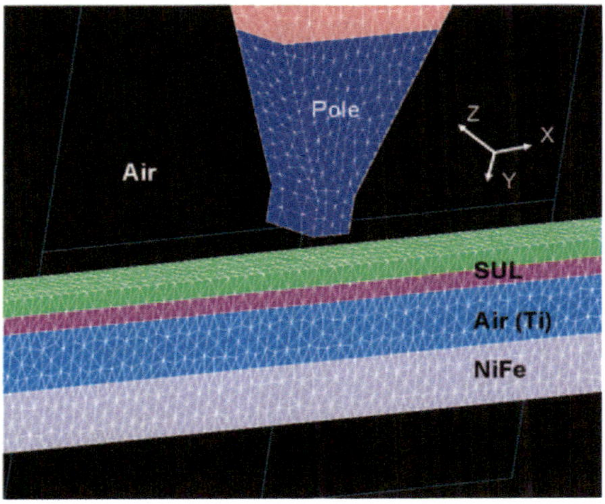

Fig. 3.32 Bird's-eye view of mesh partition of specimen for FEM calculation. Permission obtained from The Japanese Society of Microscopy [61]

3.2.4 Electron Spin Polarization

In Secs 3.1.2 and 3.2.2, the distributions of secondary electrons around the charged insulating materials were visualized by the reconstruction process of amplitude images. In this section, visualization of the electron spin polarization of secondary electrons around the charged insulating materials is presented. In the following, first detection of the magnetic field due to the electron spin polarization with a mica needle specimen under an external magnetic field was discussed, then a simpler geometrical condition was employed to conduct a detailed analysis of the distribution of

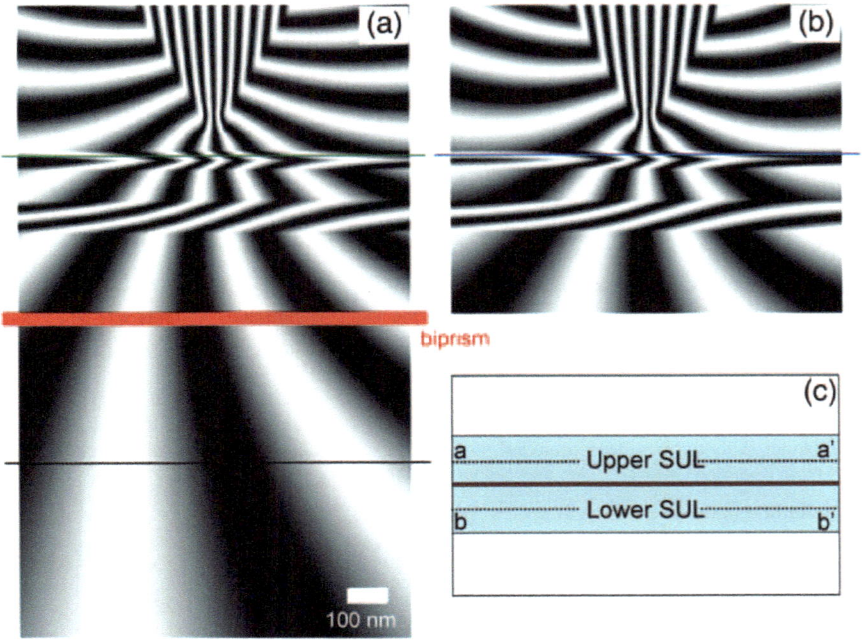

Fig. 3.33 **a** Calculated reconstructed phase image in object wave and reference wave regions. **b** Image obtained by considering the modulation of the reference wave. **c** The positions where the magnetic flux densities were quantitatively evaluated. The lines in (**a**) and (**b**) correspond to line a–a′ in (**c**). Permission obtained from The Japanese Society of Microscopy [61].

spin-polarized electrons with SiO_2 wires coated with Pt–Pd through simulation of the reconstructed phase images.

Electron Spin Polarization Around Mica Needle

The suppression of secondary-electron irradiation from a substrate was achieved by fabricating a sharp needle made from a mica specimen, as illustrated in Fig. 3.34a, using a focused ion beam (FIB) system. No appreciable changes in the distribution of secondary electrons were observed in the reconstructed amplitude images of the mica specimen under external magnetic fields of less than 20 mT. Under these conditions, the electric interactions between the surface of the positively charged mica needle and the surrounding electrons mainly determine their distribution. Unlike the experiments using a gradient in the magnetic field, such as the well-known Stern and Gerlach experiment with Ag atoms [62], a constant magnetic field across the mica needle produced spin polarization torque of electrons without a translational shift.

The magnetic interaction energy $|U_{\mathrm{m}}|$ ($\propto 1/r^3$) between spins with the dipole interaction dependent on the distance (r) in the range 0.1–10 nm is much weaker than the

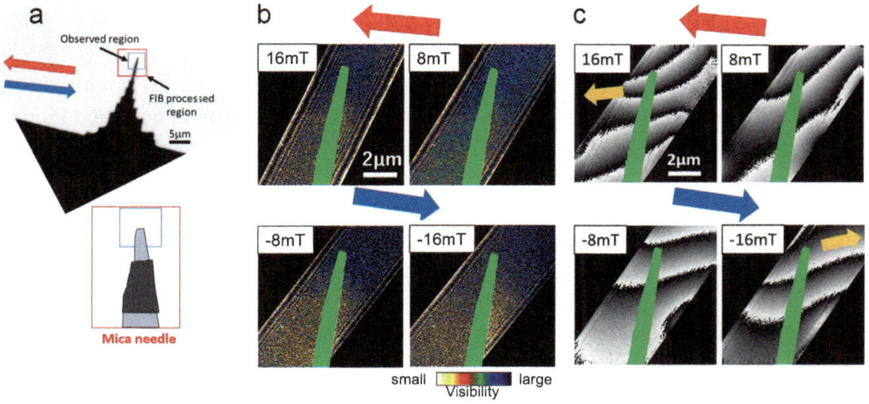

Fig. 3.34 **a** Geometric configuration of mica needle fabricated using FIB method. **b** Reconstructed phase images observed under external magnetic fields for which directions are indicated by red and blue arrows. Magnetic flux density of external magnetic field is indicated at top left corner in each image. **c** Magnetic phase information of polarized secondary electrons extracted from reconstructed phase images. Yellow arrows correspond to directions of electron spin polarization observed in reconstructed phase images. Permission obtained from Elsevier [17]

electric interaction energy U_e ($\propto 1/r$), i.e., U_e (=2.31 $\times$ 10^{-19} J) is approximately 10^7 times larger than $|U_m|$ (=1.72 $\times$ 10^{-26} J) for $r = 1$ nm [63]. Under this condition, the electric field information can easily be separated from the magnetic field information; the electric field information obtained without the external magnetic field can be subtracted from the phase information under the external magnetic field condition. The reconstructed amplitude images in Fig. 3.34b were obtained when the magnetic fields were applied along the directions indicated by the red and blue arrows. The bright yellow regions showing the distributions of secondary electrons are similarly observed in the lower part of the four images independent of applied magnetic fields. The images in Fig. 3.34c were obtained from reconstructed phase images by subtracting the electric field information, which revealed the magnetic phase information due to electron spin polarization under the external magnetic field condition.

Note that a phase image obtained in this manner can provide the difference in magnetic field information due to spin polarization between the region near the mica needle (object-wave region) and that far from the mica needle (reference-wave region). Figure 3.34c shows that the magnetic information tends to increase with the external magnetic field. It is noted that the observed magnetic information due to electron spin polarization is closely related to the distribution of secondary electrons and to the magnitude of the interaction between the electrons and the charged specimen surface [17, 64].

Electron Spin Polarization Around SiO_2 Wire

In the above study with a thin specimen of mica prepared by an FIB method, the magnetic field due to the spin polarization of secondary electrons under an external magnetic field could be detected [17]. To obtain a clear and strong signal of the magnetic field due to electron spin polarization, a simple geometric constitution containing a thin insulating wire was constructed [65]. A 0.65 μm diameter SiO_2 wire was prepared by burning quartz glass with an intense H_2–O_2 gas flame. The wire and its base were fixed to a Cu ring and coated with an approximately 30 nm thick Pt–Pd layer to suppress the severe charging effect. The geometric configuration of the magnetization stage [66] and the wire inside the stage are illustrated in Fig. 3.35. Figure 3.35a, b schematically show the top-view of the magnetization stage and its cross-section, respectively. Figure 3.35c shows an optical microscope image of the configuration of the wire and the poles of the electromagnet. Holograms were observed at the center of the wire.

In an experiment to investigate electron spin polarization using conventional electron holography, a reference wave without electric and magnetic information cannot be obtained, as noted in the previous section. The secondary electrons are widely distributed around the wire; therefore, there is an electric field due to the secondary electrons and a magnetic field due to the constant external magnetic field, and a

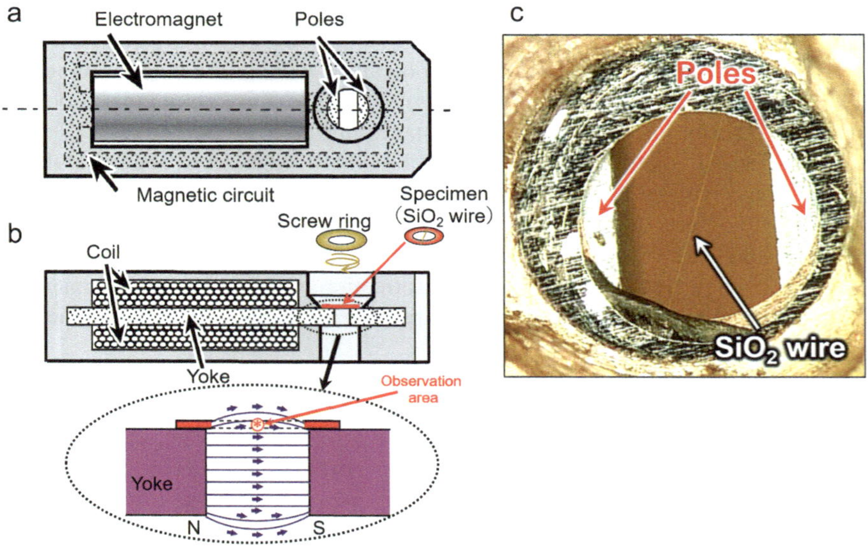

Fig. 3.35 Geometric configurations of magnetization stage in specimen holder and wire inside stage. **a** Top-view of magnetization stage at top of specimen holder. Widths of magnetic poles are 2 mm and distance between them is 1 mm. **b** Cross-section of magnetization stage. Thicknesses of poles are 1 mm. **c** Optical microscope image showing configuration of SiO_2 wire and poles of electromagnet. Permission obtained from The Japanese Society of Microscopy [65]

magnetic field due to the electron spin polarization. The external magnetic field is uniform in the reference and object wave regions, so that the phase shifts in the two regions are equal. Therefore, the uniform external magnetic field does not affect the reconstructed phase image. Although a magnetic field may be produced by the motion of secondary electrons, with and without an external magnetic field, similar motion of other secondary electrons around the charged SiO_2 wire tends to cancel this magnetic field.

Under these conditions, the electric field information can be separated from the magnetic field information, i.e., the electric field information obtained in the absence of an external magnetic field can be subtracted from the phase information obtained in the presence of an external magnetic field, which provides only information on the magnetic phase caused by electron spin polarization under the external magnetic field. In accordance with these considerations, the electron spin polarization experiment was performed as shown on the left side of the flow chart in Fig. 3.36. First, a hologram was observed without an external magnetic field, and a phase image was reconstructed from this hologram, as shown in Fig. 3.37a, which was used to separate the magnetic information from the electric information. There are irregularities in the equi-phase lines and these irregularities are attributed to the inhomogeneous charging effect of the wire. This image was also used for evaluation of the electric potential of the wire surface compared with the simulated phase image in Fig. 3.37b, which does not take into account the inhomogeneous charging effect. The distance from the wire to the magnetic poles is approximately 500 μm, which is much larger than the area where the reconstructed phase images were observed; therefore, the existence of the boundaries, i.e., magnetic poles, was neglected in the simulation of the reconstructed phase image in Fig. 3.37b. It is noted that the slope of the left side of the object region is close to the slope of the reference wave region, while the sign of the slope in the right side of the object region is opposite to that of the reference region. Therefore, the absolute values of the slopes of the phase shift in the reconstructed phase image, which is the difference of these two regions, are small in the left side and large in the right side. Through a comparison with the simulated phase image, the electric potential at the wire surface was estimated to be 0.7 V.

After a reconstructed phase image without an external magnetic field was obtained, holograms were observed for external magnetic fields (8, 14 and 20 mT), and then their reconstructed phase images were obtained. Reconstructed phase images that resulted from the phase shift due to the magnetic field of spin-polarized electrons were obtained by subtracting the phase shift due to the electric field from the phase images reconstructed from the holograms, as shown in Fig. 3.38a–c. To clarify the magnetic field distribution, color mapping is used for the phase shift. Note that the color scale is shown on the right side of Fig. 3.38c. The sign of the phase shift in the reconstructed image is assigned for the experimental condition where the specimen was observed from the incident beam direction. A reconstructed phase image obtained under an external magnetic field of 20 mT without the specimen is shown in Fig. 3.38d. The phase shift shown in Fig. 3.38d is much smaller than those in Fig. 3.38b, c. The reconstructed phase images in Fig. 3.38a–c show the differences

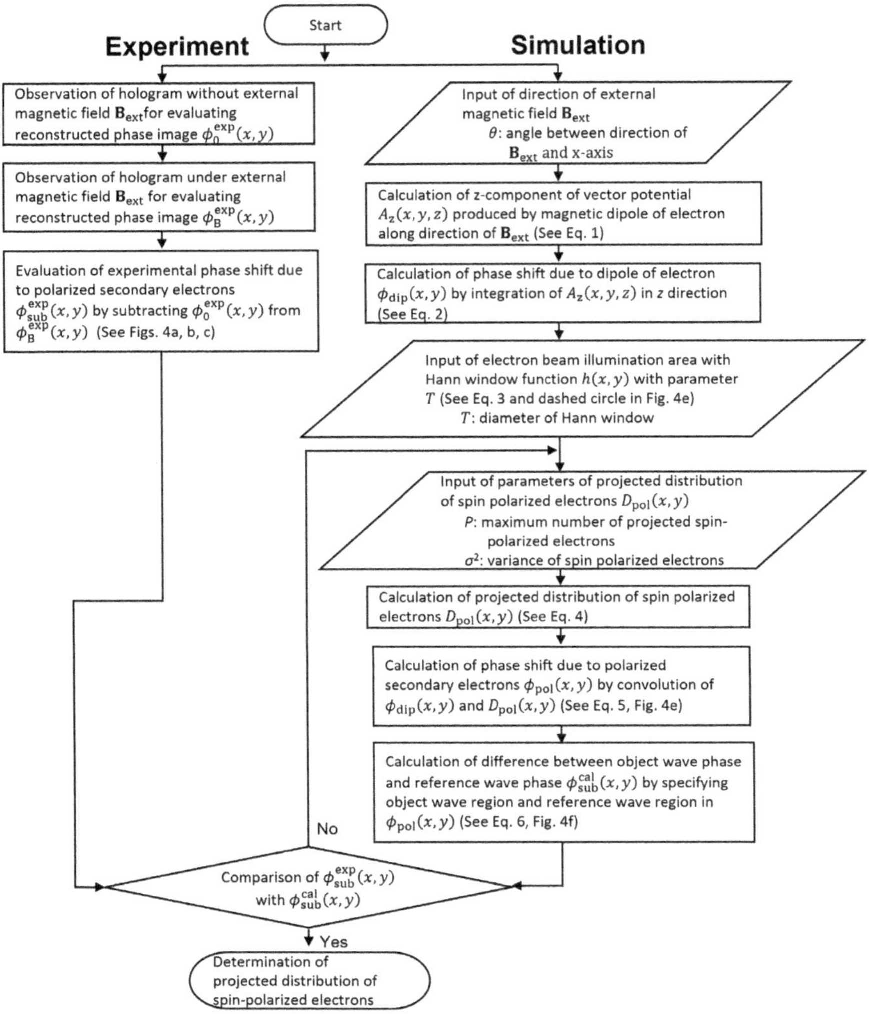

Fig. 3.36 Flow chart showing processes (experiment and simulation) for determining phase shift in reconstructed phase images. Permission obtained from The Japanese Society of Microscopy [65]

in the phase shifts due to the magnetic field of the spin-polarized electrons between the object and reference wave regions.

According to the right side of the flow chart in Fig. 3.36, simulation of the electron spin polarization was performed. The direction of the incident electrons was set parallel to the z-axis, and the wire was placed parallel to the y-axis and passing (see left side of Fig. 3.37) through the origin.

The z-component of the vector potential formed by the magnetic dipole of an electron is given [67] in SI units by

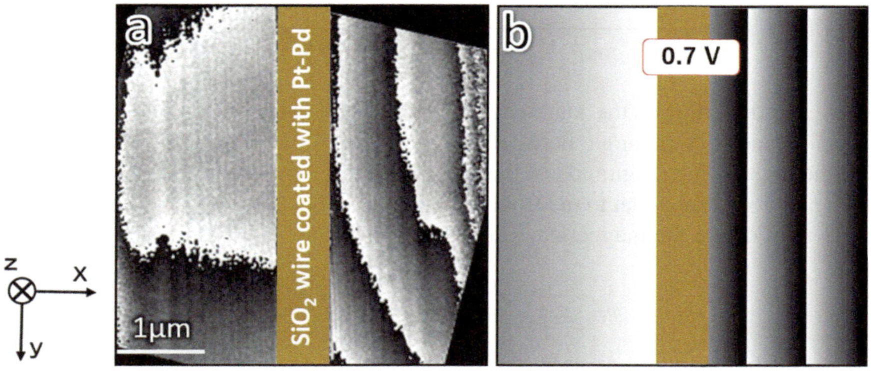

Fig. 3.37 Charging state of SiO_2 wire evaluated without external magnetic field. **a** Reconstructed phase image around wire. **b** Simulated phase image for electric potential of wire surface at 0.7 V. The phases are amplified by a factor 2. Permission obtained from The Japanese Society of Microscopy [65]

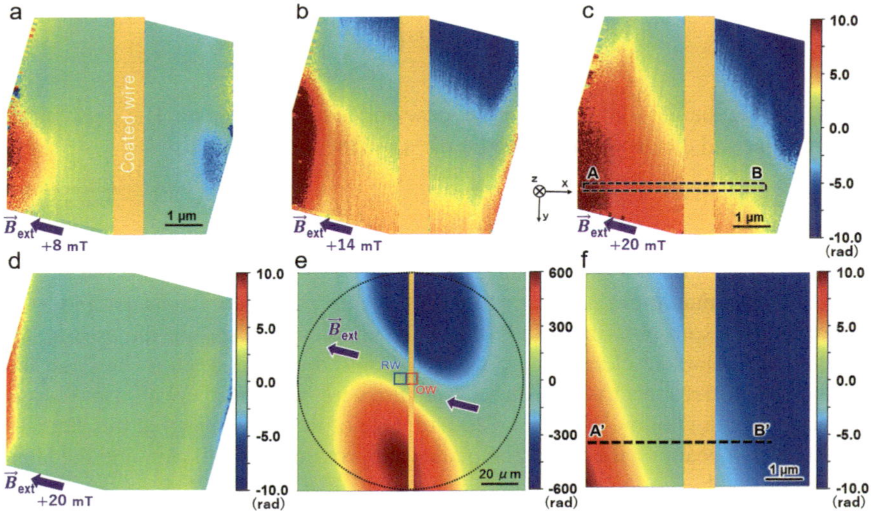

Fig. 3.38 Reconstructed phase images showing phase shift due to magnetic field of spin-polarized electrons under external magnetic fields of 8 **a**, 14 **b** and 20 **c** mT. Direction of external magnetic field is indicated by purple arrow at lower right. **d** Reconstructed phase image observed without wire specimen under external magnetic field 20 mT. **e** Simulated phase shift of incident electrons due to magnetic field of spin-polarized electrons at 20 mT. RW and OW indicate reference wave and object wave regions, respectively. Illuminating electron area is indicated by a dotted circle. **f** Simulated phase image for reconstructed phase image in (**c**). Permission obtained from The Japanese Society of Microscopy [65]

$$A_z(x, y, z) = \frac{m_d(-x\sin\theta + y\cos\theta)}{4\pi\left\{(x\cos\theta + y\sin\theta)^2 + (-x\sin\theta + y\cos\theta)^2 + z^2\right\}^{3/2}}, \quad (3.2)$$

where the direction of the magnetic dipole moment (m_d) is assumed to be in the xy-plane, and θ is the angle between the direction of the magnetic dipole moment and the x-axis. The magnitude of the magnetic moment due to the magnetic dipole of an electron is set equal to the Bohr magneton (9.2740×10^{-24} J/T); therefore, the phase shift due to the magnetic moment is given by

$$\phi_{dip}(x, y) = -\frac{2\pi e}{h} \int A_z(x, y, z)dz, \quad (3.3)$$

where it was simply assumed that a spin-polarized electron has a magnitude of the magnetic moment that corresponds to the Bohr magneton, due to precession in the direction of the applied external magnetic field [68]. The Hann window function was used to construct the distribution of spin-polarized electrons:

$$h(x, y) = 0.5 + 0.5\cos\left(\frac{2\pi\sqrt{x^2 + y^2}}{T}\right), \quad (3.4)$$

where T is the diameter of the Hann window and is set to 120 μm. The distribution of spin-polarized electrons projected on the xy-plane is obtained as

$$D_{pol}(x, y) = h(x, y) \cdot p \cdot \exp\left(-\frac{x^2}{2\sigma^2}\right), \quad (3.5)$$

where p is the maximum number of projected spin-polarized electrons, and σ^2 is the variance of spin-polarized electrons. This equation is based on the assumption that the spin-polarized electrons are projected onto the xy-plane in a Gaussian-like distribution with respect to the distance from the wire. In (3.5), there are two reasons for using a Hann window with $T = 120$ μm. One of the reasons is that the diameter of the electron beam on the specimen plane is approximately 120 μm, and the Hann window limits the area where polarized secondary electrons are mainly distributed. The other reason is to reduce the artifact of the boundary condition during the convolution operation when the phase shift due to all the polarized secondary electrons ϕ_{pol} is estimated. ϕ_{pol} is evaluated by convolution of the phase shift due to a dipole ϕ_{dip} and the projected distribution of polarized secondary electrons D_{pol}:

$$\phi_{pol}(x, y) = \phi_{dip}(x, y) \otimes D_{pol}(x, y), \quad (3.6)$$

where $\otimes$ is the convolutional operator. Finally, the phase shift of the reconstructed phase images is obtained from the difference between the phase shifts of the object wave and the reference wave:

$$\begin{aligned}\phi_{\mathrm{sub}}^{\mathrm{cal}}(x, y) &= \phi_{\mathrm{obj}}(x, y) - \phi_{\mathrm{ref}}(x, y) \\ &= \phi_{\mathrm{pol}}(x, y) - \phi_{\mathrm{pol}}(x - w, y),\end{aligned} \tag{3.7}$$

where w is the width of the interference region where the object and reference waves overlap.

As shown on the right side in Fig. 3.36, the reconstructed phase image (Fig. 3.38c, 20 mT) was investigated by computer simulation. Figure 3.38e shows the phase shift distribution of incident electrons due to the magnetic field of spin-polarized electrons at 20 mT. The values of the parameters p and σ in (3.5) used to obtain this phase shift distribution are 2.54 $(100\ \mathrm{nm}^2)^{-1}$ and 20 μm, respectively. The dotted circle corresponds to the area of incident electron irradiation. At the center of the irradiated area, the intensity of the incident electrons is the highest and it decreases towards the dotted circle. Thus, the density of secondary electrons is considered to be the highest at the central part of the irradiated area. Note that the increase in the phase shift from the blue region at the upper right to the red region in the lower left is perpendicular to the direction of the applied magnetic field.

In Fig. 3.38e, the phase shift distribution shows that the direction of the magnetic flux is parallel to the applied magnetic field direction at the center of the wire region and that the magnetic flux spreads out and turns around, reversing its direction along the dotted line corresponding to the electron irradiation area. The phase shift distribution does not have line symmetry along the line direction of the external magnetic field indicated by the purple arrows due to the distribution of secondary electrons along the wire.

Subtraction of the phase shift of the reference wave (RW) region from that of the object wave (OW) region gives a simulated phase image shown in Fig. 3.38f. The color change that corresponds to the phase shift due to the magnetic field caused by the spin-polarized electrons from the blue region in the upper right to the red region in the lower left is basically in good agreement with the color change in Fig. 3.38c.

The distribution of spin-polarized electrons projected along the incident beam direction is shown in Fig. 3.39, which was used for the simulation of Fig. 3.38e. The number of projected spin-polarized electrons is the largest in the central part of the wire region and decreases with an increase in distance from the wire.

The reconstructed phase image in Fig. 3.38f is the difference between the phase shifts of the object wave and the reference wave, as indicated by $\phi_{\mathrm{sub}}^{\mathrm{cal}}$ in (3.7). The slope of this phase distribution $\phi_{\mathrm{sub}}^{\mathrm{cal}}$ corresponds to the distribution $\mathbf{B}_{\mathrm{sub}}(x, y)$, which is the projected distribution of the magnetic flux density in the object wave region minus that of the magnetic flux density in the reference wave region. The projected magnetic flux densities due to the spin-polarized electrons at the boundaries of the object wave region, which are the same distance from the wire, are almost equal,

$$\mathbf{B}\left(\frac{w}{2}, y\right) \approx \mathbf{B}\left(-\frac{w}{2}, y\right). \tag{3.8}$$

Note that the right edge of the reference wave region and the left edge of the object wave region are the same. Therefore, the following relation is obtained, i.e.,

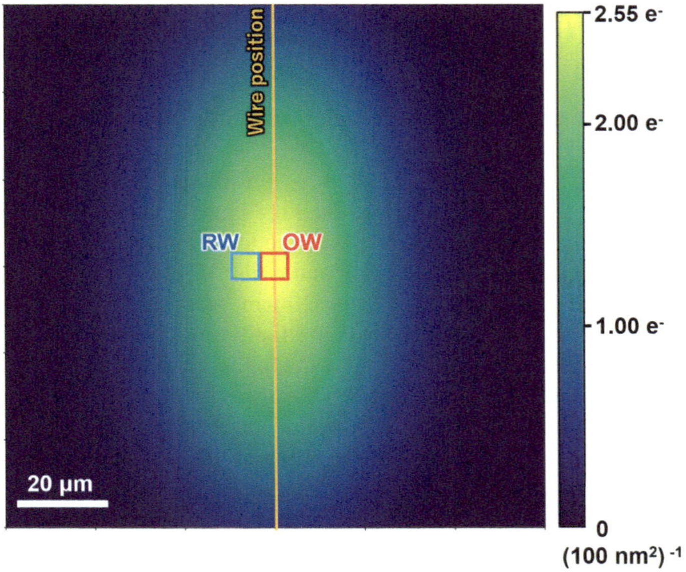

Fig. 3.39 Projected distribution of spin-polarized electrons around charged SiO_2 wire. Permission obtained from The Japanese Society of Microscopy [65]

$$\mathbf{B}_{\mathrm{sub}}\left(\frac{w}{2}, y\right) \approx 0. \tag{3.9}$$

There is also the following relationship between the phase slope and the magnetic flux density in the reconstructed phase images:

$$\left(\frac{\partial \phi_{\mathrm{sub}}}{\partial x}, \frac{\partial \phi_{\mathrm{sub}}}{\partial y}\right) = \left(-\frac{2\pi e B_{\mathrm{sub\_y}}}{h}, \frac{2\pi e B_{\mathrm{sub\_x}}}{h}\right). \tag{3.10}$$

Thus, substituting (3.9) into (3.10) yields the following relation,

$$\frac{\partial \phi_{\mathrm{sub}}\left(\frac{w}{2}, y\right)}{\partial x} \approx 0. \tag{3.11}$$

The phase profile of the observed phase image in Fig. 3.38c between A and B is presented in Fig. 3.40a, while the phase profile of the simulated phase image in Fig. 3.38f between A' and B' is presented in Fig. 3.40b. The observed phase image in Fig. 3.38c contains the fluctuation of data and the shape of the image is not a square; therefore, a band area was selected in the lower left, and the data were averaged along the y-direction, as shown in Fig. 3.40a. In Fig. 3.40b, the phase profile of the simulation at the corresponding region between A' and B' is shown. As indicated by the dotted lines that were extrapolated along the phase profiles, the absolute values of the slopes in both phase profiles are large at regions A and A', and these values

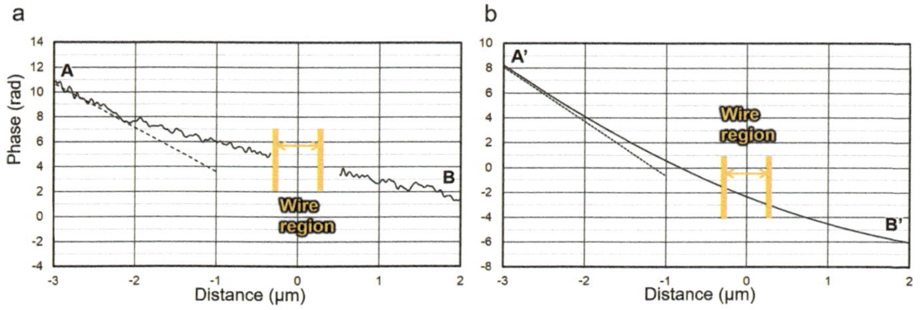

Fig. 3.40 Slopes of phase profiles in reconstructed phase images. **a** Observed slope of phase profile along A–B in Fig. 3.38c. **b** Simulated slope of phase profile along A'-B' in Fig. 3.38f. Permission obtained from The Japanese Society of Microscopy [65]

become smaller toward B and B' in Fig. 3.40a, b, respectively. These results confirm the evidence of the phase shift due to the spin-polarized electrons.

Reconstructed phase images were obtained with the same experimental procedure under external magnetic fields of −8, −14 and −20 mT as that in Fig. 3.38, as shown in Fig. 3.41a–c, respectively. A reconstructed phase image obtained without the SiO_2 wire under an external magnetic field −20 mT is shown in Fig. 3.41d. Note that the phase shift of Fig. 3.41d is much smaller than those of Fig. 3.41b, c. Figure 3.41e shows the phase shift distribution of incident electrons due to the magnetic field of spin-polarized electrons at −20 mT. The values of p and σ in (3.5) are 1.89 (100 $nm^2)^{-1}$ and 20 μm, respectively. Note that in contrast to Fig. 3.38e, the increase in the phase shift from the blue region at the lower left to the red region in the upper right is perpendicular to the direction of the applied magnetic field.

Subtraction of the phase shift in the reference wave (RW) region from that in the object wave (OW) region in Fig. 3.41e gave a reconstructed phase image for the simulation in Fig. 3.41f. The color change corresponding to the phase shift due to the magnetic field caused by the spin-polarized electrons from the blue region in the lower left to the red region in the upper right is in good agreement with the color change in Fig. 3.41c.

The present experimental results are interpreted in relation to the magnetic moment of the spin-polarized electrons under an external magnetic field, as follows. We consider the motion of secondary electrons with two energy regions in the stationary state. The first is the energy region less than approximately 0.7 eV. Here, the magnetic Lorentz force and the electric force near the charged SiO_2 wire are compared. The magnetic Lorentz force on a secondary electron is given by $|\mathbf{F}_{mag}|=|\mathbf{v}\times\mathbf{B}\,e| \leq vBe$ (v: velocity of secondary electron; B: magnetic flux density; e: elementary electric charge). If we assume the energy of a secondary electron to be 0.3 eV whose velocity is *ca.* 3.25×10^5 m s^{-1} and the external magnetic flux density to be 20 mT, $|\mathbf{F}_{mag}(N)|$ is less than approximately 3.3×10^5 (m s^{-1}) $\times$ 20 $\times$ 10^{-3}(T) $\times$ e (C) = 6.6×10^3(N/C) $\times$ e (C).

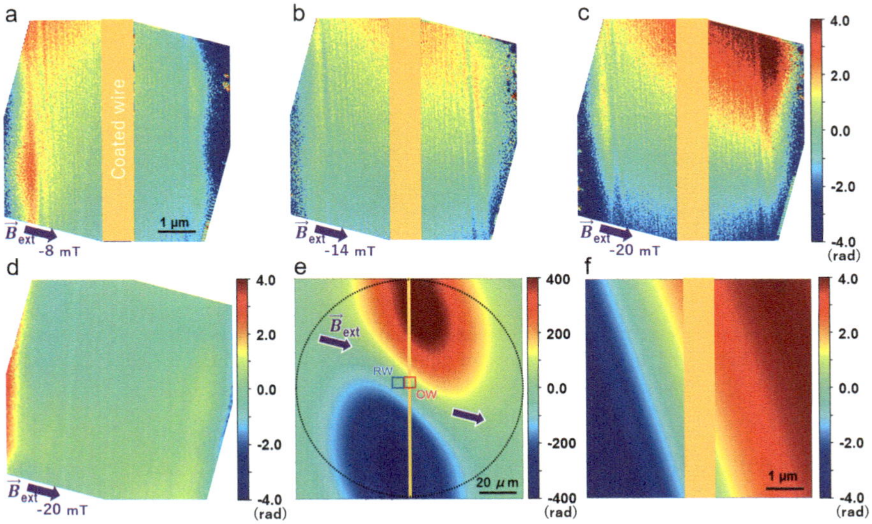

Fig. 3.41 Reconstructed phase images showing phase shift due to magnetic field of spin-polarized electrons under external magnetic fields of −8 (**a**), −14 (**b**) and −20 (**c**) mT. Direction of external magnetic field is indicated by purple arrow at lower right. **d** Reconstructed phase image observed without wire specimen under external magnetic field −20 mT. **e** Simulated phase shift of incident electrons due to magnetic field of spin-polarized electrons at −20 mT. RW and OW indicate reference wave and object wave regions, respectively. Illuminating electron area is indicated by a dotted circle. **f** Simulated phase image for reconstructed phase image in (**c**). Permission obtained from The Japanese Society of Microscopy [65]

The magnitude of the electric force on a secondary electron due to the charged wire is given by $|\mathbf{F}_{el}|=eE$ (E: electric field of the wire). The electric field at the position 0.4–0.5 μm distant from the charged SiO_2 wire surface (0.7 V) is approximately 30 mV/100 nm, and $|\mathbf{F}_{el}(N)|$ is approximately 3×10^5(N/C) × e(C). Therefore, the electric force on the secondary electrons is much larger than the magnetic Lorentz force.

In Fig. 3.42a, secondary electrons are assumed to leave the position 0.4 μm apart from the center of the wire at position ①. Although secondary electrons are emitted in various directions in real situations, to understand the basic features, it is simply assumed that the emission direction of secondary electrons is perpendicular to the wire on the basis of the wire symmetry. For secondary electrons with energies less than approximately 0.7 eV, the motion of the secondary electrons is mainly controlled by the attractive electric force from the charged wire. In the simulation, the electric potential is assumed to be zero at the distance of the magnetic poles, i.e., about 500 μm from the wire. A secondary electron, for example, with a kinetic energy of 0.5 eV loses the kinetic energy and gains potential energy. The secondary electron eventually reaches position ②, which is approximately 120 μm apart from the wire (kinetic energy approximately zero and potential energy approximately maximum), and then the secondary electron comes back near the wire through position ③ with an increase in its kinetic energy. It takes approximately 2.3 ns to reach position ① again. The reaching distance and the round trip time for electrons with energies less than 0.6 eV are presented in Fig. 3.42b. More or less secondary electrons with much lower energy are present near the wire; therefore, the secondary electron then undergoes the repulsive force that interacts with these lower energy electrons. Then, the secondary electron tends to leave the wire as indicated by the dotted arrow ④, and then undergoes the attractive force again from the charged wire, which results in the repeated motion of leaving the wire and coming to the wire. The secondary electron loses the energy through such motion due to the inelastic scattering processes, and then the reaching distance from the wire becomes gradually smaller. Finally, the secondary electron with much lower energy comes near the coating layer and moves to the ground.

Secondary electrons with higher energies more than approximately 0.7 eV can also be emitted from the wire in various directions. These secondary electrons are affected mainly by Lorentz magnetic forces. If it is assumed that the emission direction is perpendicular to the external magnetic field of 20 mT, then a simple circular motion tends to be formed. The radius of the circular motion is listed in Table 3.2 as a function of the initial energy, and the time for the circular motion is constant ($2\pi m/eB = 1.79$ ns). When these secondary electrons come near the wire, they interact with secondary electrons with much lower energies and lose their kinetic energy. These secondary electrons then behave similarly as the first case and tend to lose their energies, which results in the movement to the ground through the coating layer.

In both cases noted above, the generation of new secondary electrons with relatively high energy and the disappearance of secondary electrons with much lower energies that moves through the coating layer are in balance for the stationary state. Here it should be noted that this balance is satisfied through the energy loss processes

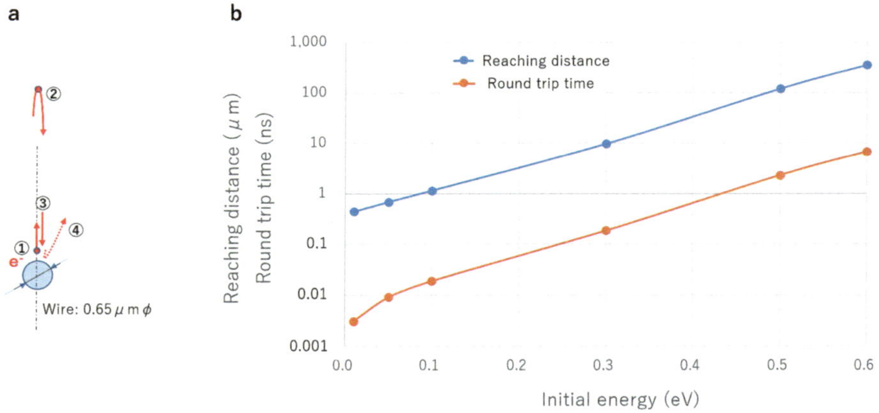

Fig. 3.42 **a** Schematic illustration showing motion of secondary electron around charged wire. **b** Reaching distance and round trip time as function of initial energy of secondary electron. Permission obtained from The Japanese Society of Microscopy [65]

Table 3.2 Velocity and radius of circular motion of secondary electron for energy 1–10 eV

E (eV)	v (m/s)	r (μm)
1.0	5.9309×10^5	169.0
3.0	1.0273×10^6	292.0
5.0	1.3262×10^6	377.0
10.0	1.8755×10^6	533.0

of newly produced secondary electrons. Therefore, in the stationary state, the probability of the inelastic scattering process of secondary electrons that lose their energy is higher than that of a gain in energy.

Through these processes of secondary electrons in the stationary state distributing around the wire for a much longer time than the time scale of nanoseconds for spin flipping, as one of such energy loss processes, the secondary electrons eventually tend to take a more stable spin direction parallel to the external magnetic field. Additional explanations about the electron spin polarization are as follows.

Electrons have the spin angular momentum $\overrightarrow{S}$, of which the magnitude is $S = \sqrt{3}\hbar/2$. The magnitude of the spin magnetic moment of electrons is given by $g_s\mu_B S/\hbar$, where $g_s$is the spin g-factor ($\approx$2) and $\mu_B (= e\hbar/2m)$ is the Bohr magneton. Under an external magnetic field, electrons undergo Larmor precession with the axis of rotation in the direction of the external magnetic field. The quantum mechanical expected values of the spin magnetic moment are $+g_s\mu_B s$ and $-g_s\mu_B s$ ($s = 1/2$: spin quantum number), which are parallel and antiparallel to the axis of rotation in the direction of the external magnetic field, respectively. The Larmor angular velocity $\omega_s(=2\pi\nu_s)$ is given by $g_s\mu_B B_{ext}/\hbar$ and the frequency ν_s for B_{ext} (=20 mT) is 5.6 $\times$ 10^8 s^{-1}. The energy change ($\Delta E = 2g_s\mu_B s B_{ext}$) between the two states of electron spin is a small value, and this energy change is around 2.3 $\times$ 10^{-6} eV for B_{ext} (=20 mT). The state of the electron spin parallel to the external magnetic field is energetically stable than the other state of the antiparallel electron spin; therefore, the probability of taking the spin state parallel to the external magnetic field becomes larger, although in the initial state, the probabilities of secondary electrons with quantum mechanical expected values of the spin magnetic moments $+g_s\mu_B s$ and $-g_s\mu_B s$ are equal. Note that the net distribution of spin-polarized electrons parallel to the external magnetic field minus the antiparallel spin-polarized electrons projected along the incident electron beam during the exposure time (6 s) is shown in Fig. 3.39. The simulation of the net distribution of the phase shift due to the magnetic moment of the spin state parallel to the external magnetic field minus that of antiparallel spin state projected along the incident electron beam during this exposure time is shown in Figs. 3.38e and 3.41e. Further detailed holography investigations are expected to clarify the distribution of spin-polarized electrons in relation to the charged state and the surface condition of various insulating material surfaces [64].

3.3 Plates

Plate-shape specimens are widely utilized for electron microscope observation. In addition to evaporated films, thinned specimens with a plate-shape prepared by the FIB method can be extensively utilized.

3.3.1 Electric Field

Diamond-Like Carbon

Figure 3.43a shows a cross-sectional view of a diamond-like carbon (DLC) film deposited onto a Si substrate [69]. The film thickness can be directly determined from the cross-sectional view. Figure 3.43b shows an electron hologram of a DLC film after it was removed from a Si substrate by chemical etching. The dark dotted line outlines the specimen edge. Figure 3.43c shows the phase shift measured along line X (in vacuum) and line Y (crossing both the vacuum and the specimen), as indicated in Fig. 3.43b. The fringes were scanned in the direction of the white arrow shown in Fig. 3.43b. The peak shifts of interference fringes were plotted as a function of the peak number. Analysis of the phase shift in the DLC film indicated the inner potential was 10.8 ± 0.9 V. When the mean inner potential is known, (3.1) can be used to determine the thickness of the fabricated film. This method has been demonstrated to be applicable to DLC films as thin as 5 nm.

Oxide Heterojunctions

In oxide heterojunctions, spontaneous polarization plays a dominant role in the charge screening. The materials studied by electron holography are trivalent transition-metal-oxide insulators, which do not form a conductive interface with $SrTiO_3$ (STO) even though the interface has a polar discontinuity [70]. $LaFeO_3$ (LFO) is known as a centrosymmetric Mott insulator with an optical gap (2.2 eV) in the visible-light region [71]. Because of the closed-shell structure in the up-spin band of the Fe^{3+} state, LFO exhibits a high resistivity compared with other Mott insulators [72]. LFO also exhibits instability toward the polar state under epitaxial strain [73]. Thus, the LFO = STO junction is an attractive model system to examine interface-induced spontaneous polarization. 30 nm thick LFO films were deposited on 0.5 mm thick STO and $SrTi_{0.9998}Nb_{0.0002}O_3$ (Nb:STO) substrates with controlled TiO_2 and SrO terminations, as illustrated in the left side of Fig. 3.44a [74].

Figure 3.44a shows the atomic structure and elemental distributions near the LFO = STO interfaces imaged by scanning transmission electron microscopy (STEM) and energy-dispersive X-ray spectroscopy (EDS). The image shows that the elemental

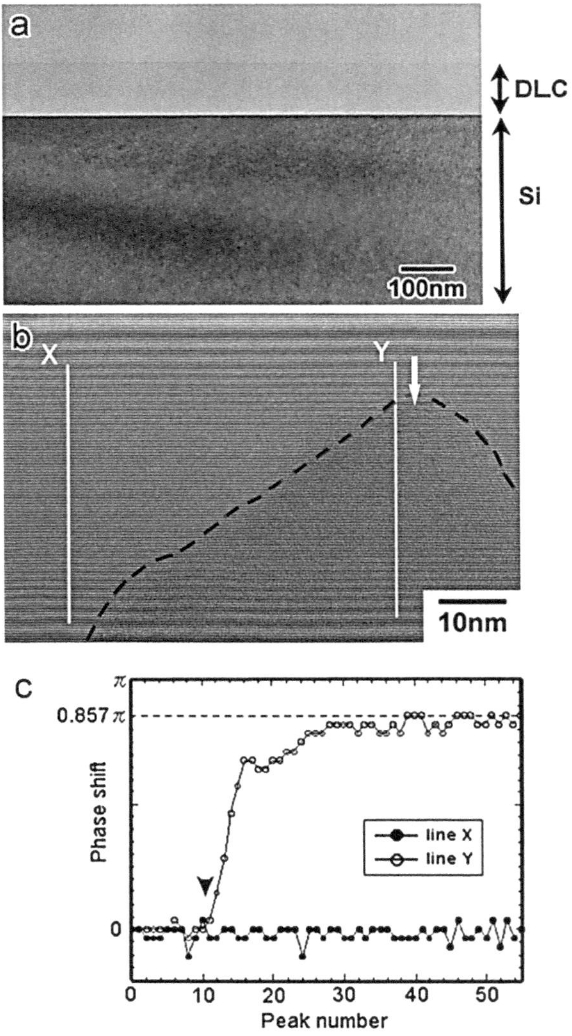

Fig. 3.43 **a** Cross-sectional view of a DLC film. **b** Electron hologram of the DLC film. Dotted line outlines the specimen edge. (c) Phase shift observed along lines X and Y indicated in (**b**). Permission obtained from Elsevier [17]

sequence was constructed as designed. From the STEM–EDS analysis, the broadening of the distribution was estimated at most 2 unit cells from the interface [74], which indicates that the interface is sufficiently abrupt for the interfacial atomic sequence to be classified into two definite types and that the effect of interfacial disorder is negligible. The absence of lateral conductivity in the LFO = STO junctions, irrespective of the interface terminations, was confirmed. Electron holography can directly quantify the spatial distribution of the electrostatic potential and, accordingly, the built-in internal electric field in a specimen [74]. Mappings of the electrostatic potentials for an electron is shown in Fig. 3.44b, where the junctions formed on

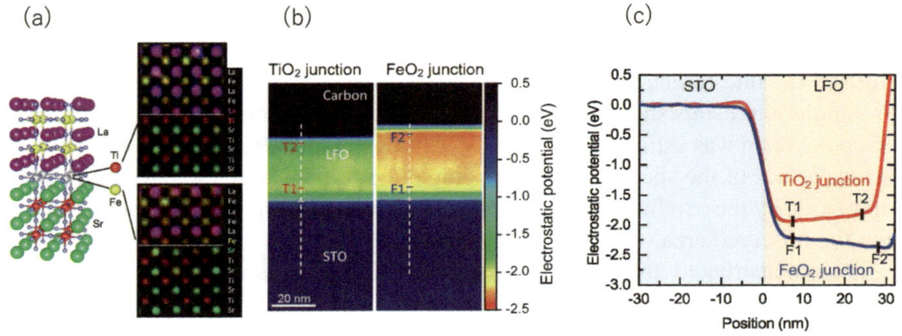

Fig. 3.44 **a** Elemental distribution mappings for a TiO_2 junction (upper panel) and FeO_2 junction (lower panel) taken by STEM, where atomic-resolution EDS images are superimposed on HAADF images. **b** Mappings of the electrostatic potentials derived from electron holography for two LFO/STO junctions with different interface terminations. **c** Line profiles of the potentials along the dashed lines in (**b**). Permission obtained from American Physical Society [74]

the TiO_2- and SrO-terminated STO surfaces are denoted as the TiO_2 junction and the FeO_2 junction according to the interfacial transition-metal layer, respectively. The electrostatic potential is almost uniform in the in-plane direction in both junctions. The electrostatic potential can now be examined in detail; Fig. 3.44c shows two examples of the electrostatic potential obtained along the dashed lines in Fig. 3.44b.

The spontaneous polarization dominates the electric polarization as shown by electron holography. The polarization direction can be switched by changing the chosen interfacial atomic sequence. Note that the emergent spontaneous polarization induces anomalous photovoltaic properties in the junctions, such as a sign reversal in the photocurrent that is dependent on the interfacial atomic sequence and the excitation photon energy.

3.3.2 *Electric Field Fluctuation Due to the Collective Motion of Electrons*

In the work described in Sects. 3.1.2 and 3.2.2, the collective motion of electrons was studied with specimens of particles and lines, respectively. In this section, the dependency of secondary-electron accumulation on the surface condition of charged insulators is studied with a plate specimen by utilizing the specimen techniques, i.e., ultramicrotomy and an FIB method. The specimen is an epoxy resin, which is a rather soft material that can be sliced by ultramicrotomy.

First, an epoxy specimen is prepared by ultramicrotomy to provide a clear specimen surface devoid of metallic elements. Figure 3.45a shows a reconstructed phase image of a thin epoxy resin film (dark brown region) prepared by ultramicrotomy. A reconstructed phase image was simulated to estimate the electric potential of the top surface of the specimen. The thickness of the model was 2.0 μm, and the electric potentials on the sides and bottom of the specimen surfaces were set to 0.0 V. The width of the interference fringe region was assumed to be 2.8 μm. Compared with the simulation results shown in Fig. 3.45b, the electric potential of the top surface of the epoxy resin was estimated to be 1.2 V. Although there were no metallic elements on the surface of the specimen, the charging effect is considered to have been partly suppressed by the irradiation of secondary electrons from the specimen support plate near the observed area.

The reconstructed amplitude image in Fig. 3.45c shows red regions around the surface of the epoxy resin, which are considered to correspond to high densities of secondary electrons that strongly interact with the surface of the positively charged specimen. Particularly in the concave region, as indicated by the arrow in Fig. 3.45c, a bright red region is evident, which corresponds to a large electric field fluctuation due to interaction of the accumulated secondary electrons with the surface, similar to the case of the branched microfibril in Fig. 3.27.

After observation of the epoxy resin hologram, both sides of the thin specimen were irradiated using an FIB system with a weak Ga-ion beam. The beam intensity was 0.85×10^3 mC·m^{-2}, which is 200 times less than that usually used to polish specimens. In Fig. 3.45d, the reconstructed phase image shows that the electric potential of the specimen is 1.0 V by comparison with the simulated phase image shown in Fig. 3.45e. Because of the weak Ga-ion beam, the electric potential of the

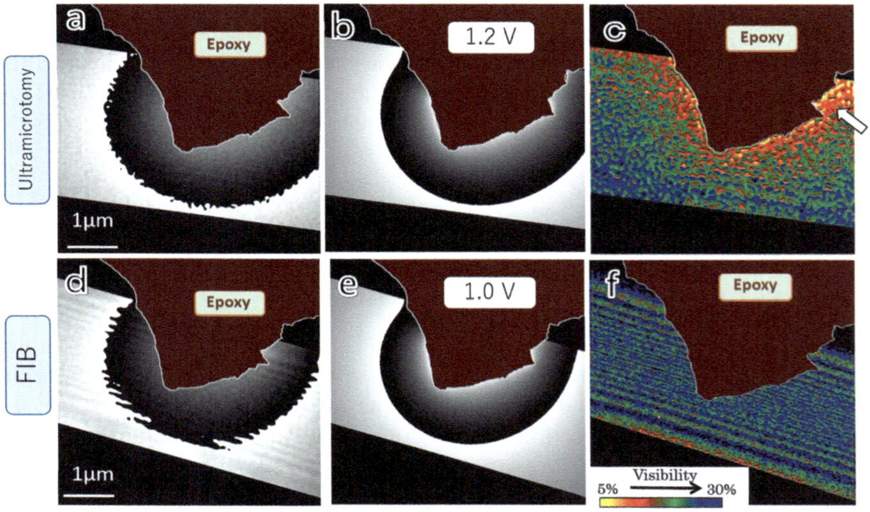

Fig. 3.45 **a** Reconstructed phase image of ultramicrotomed epoxy resin. **b** Simulated phase image with surface electric potential of 1.2 V. **c** Reconstructed amplitude image of ultramicrotomed epoxy resin. **d** Reconstructed phase image of epoxy resin treated with FIB system. **e** Simulated phase image with electric potential of 1.0 V. **f** Reconstructed amplitude image of epoxy resin treated with FIB. Permission obtained from Elsevier [17]

specimen before and after irradiation does not differ significantly. Note that the shape of the specimen in Fig. 3.45d is almost the same as that in Fig. 3.45a, which indicates that the irradiation damage caused by the Ga ions was negligible.

In the reconstructed amplitude image in Fig. 3.45f, the red regions evident in Fig. 3.45c are no longer visible. Therefore, the distribution of secondary electrons that interact strongly with the surface of the positively charged epoxy resin is quite sensitive to the presence of metallic elements on the surface [16].

3.3.3 Magnetic Field

Magnetization Process of Mn-Zn Ferrite

Soft ferrites (e.g., Mn–Zn ferrite, which is widely used in transformers) are a typical soft magnetic material [75, 76]. A magnetization process with an external magnetic field produced by an electromagnet (Figs 3.46a, b) was investigated to understand the low coercivity of these materials.

Figure 3.47 shows the change in the reconstructed phase images of Mn–Zn ferrite with an increase of the applied magnetic field [77]. When these reconstructed phase images were observed, the misalignment of the electronic optical axis and astigmatism were corrected. When the applied magnetic field was smaller than 1600 A m^{-1}, no change was observed in the distribution of the magnetic flux (Fig. 3.47a, b). However, the distribution of the magnetic flux changed from the state shown in Fig. 3.47b to that shown in Fig. 3.47c when the magnetic field was increased to 1760 A m^{-1} (refer to the domain-wall motion). The magnetic flux in Fig. 3.47d changes preferentially at the grain boundary and a new domain wall, as indicated by the black dashed line, appears at this boundary. This phenomenon appears to be due to the decrease in permeability at a grain boundary that includes precipitates such as SiO_2. However, a new domain wall, as indicated by the white dashed line, appears in Fig. 3.47f. This domain wall has moved to the right as a result of the increase in the intensity of the applied magnetic field. Finally, the lines of magnetic flux have the same direction as that of the applied magnetic field on the right side of the image in Fig. 3.47i. In contrast, an antiparallel flux is observed on the left side, which was likely caused by a thick area formed during the FIB milling process (see the video (cutoff animation) in Fig. 3.48).

Domain Wall Motion in Electrical Steel Sheets

Lorentz microscope observations of domain walls in soft magnetic materials can be performed in situ under an alternating current (AC) magnetic field using an AC magnetic system (Fig. 3.46c). Since the exposure time to observe holograms is around several seconds and interference fringes are sensitive to the fluctuation of the irradiation angle due to the applied external magnetic field, dynamic observation of domain-wall motions by electron holography is difficult. On the other hand, Lorentz microscopy with a Fresnel mode [78–83] is useful, as shown below. Gründmayer and Zweck [84] used a thin metallic wire to apply an AC magnetic field. The use of a metallic wire was reported earlier by Yi et al. [80], although they achieved a pulse magnetic field rather than an AC magnetic field. Akase et al. used a small electromagnet to apply an AC magnetic field [82] to an electrical steel sheet [85] and nanocrystalline soft magnetic $Fe_{85}Si_2B_8P_4Cu_1$ alloy [86].

Lorentz micrographs with Fresnel mode of a doubly oriented electrical steel sheet in an AC magnetic field that were captured from a videotape as shown in Fig. 3.49 [85]. The frequency and amplitude of the AC magnetic field were 0.34 Hz and 2.5 kA m^{-1}, respectively. The frames from Fig. 3.49a–d correspond to one-half of one AC cycle. The left column shows the original micrographs; the dashed lines and small arrows in the micrographs on the right indicate the domain-wall contrast and magnetization, respectively. In addition, the arrowhead indicates a SiO_2 precipitate and the large black arrow at the top-right indicates the direction of the applied magnetic field. The domain walls are observed to move smoothly from Fig. 3.49a–d; therefore, the area of the domain whose magnetization is parallel to the applied magnetic field increases with magnetic field. On the other hand, the domain with magnetization antiparallel to the applied field disappears in Fig. 3.49d. It is noted that the domain walls tend to have a linear shape parallel to the axis of easy magnetization (see a video of the dynamic motion of domain walls in Fig. 3.50).

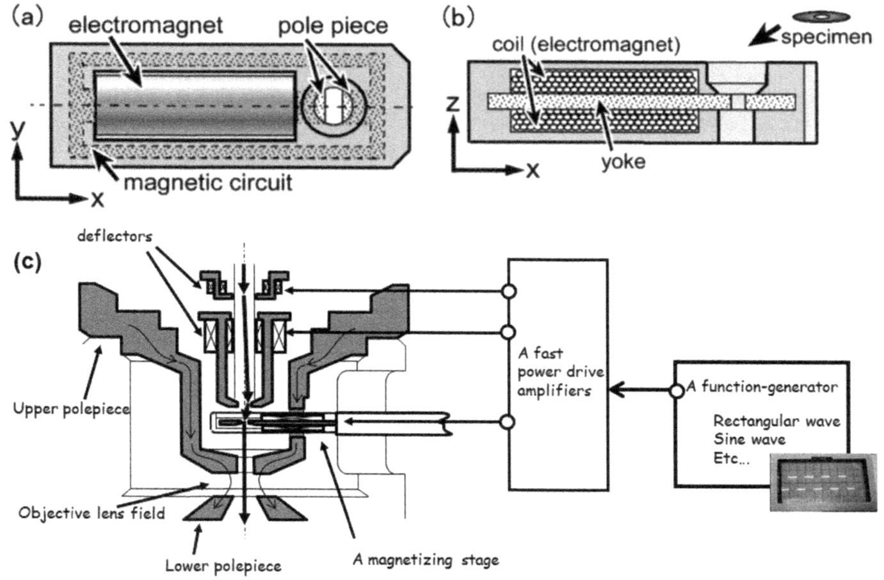

Fig. 3.46 Schematic of head of a magnetizing stage: **a** a plane view and **b** a cross-sectional view. **c** Schematic of a magnetizing system for in situ Lorentz microscopy using an AC magnetic field. Permission obtained from Elsevier [17]

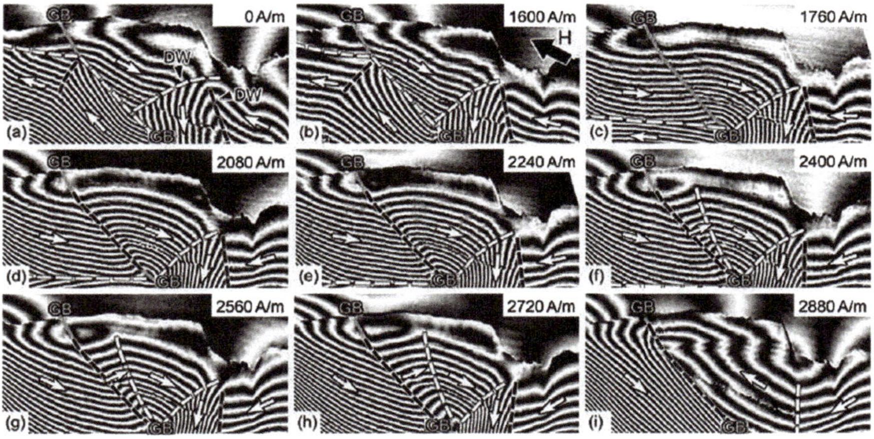

Fig. 3.47 **a–i** Change with increasing magnetic field in reconstructed phase images of a Mn–Zn ferrite specimen. White arrows indicate directions of lines of magnetic flux. Black arrow in **b** represents direction of applied magnetic field. Broken lines indicate positions of magnetic domain walls (DWs). Line between "GB" indicates position of a grain boundary. Permission obtained from Elsevier [17]

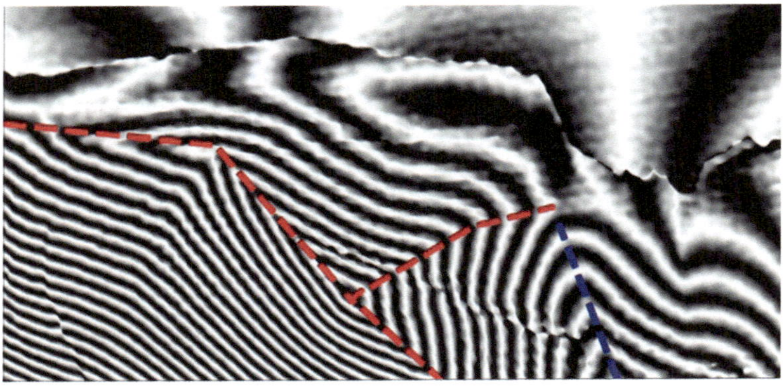

Fig. 3.48 Video: Cutoff animation of magnetization process in Mn–Zn ferrite

Figure 3.51 shows Lorentz micrographs of a non-oriented electrical steel sheet observed under the two-beam condition [85]. The contrast of the strain fields around the precipitates and many dislocations is enhanced showing the positions of these defects. When an AC magnetic field (sine wave, 1.0 Hz, 2.4 kA m^{-1}) is applied, the domain walls are trapped at the precipitates (Fig. 3.51a) and at the regions around their strain fields (Fig. 3.51b). Dynamic observation of the interactions between domain walls and lattice defects is important to understand eddy-current and hysteresis losses in electrical steel sheets. The inclusions have been introduced to improve the mechanical properties. As shown above, domain walls are pinned at these inclusions and the

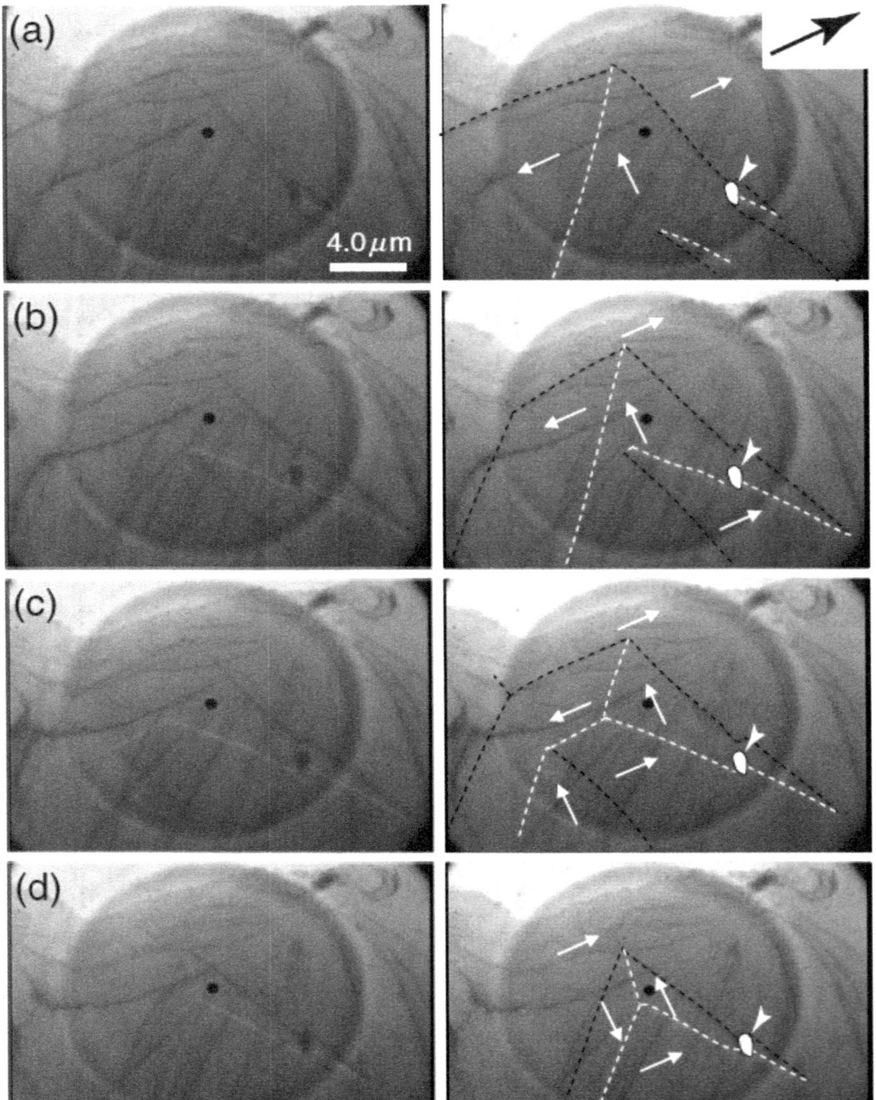

Fig. 3.49 Lorentz micrograph of doubly oriented electrical steel sheet in an AC magnetic field (0.34 Hz, 2.5 kA m^{-1}), as captured from a videotape. Large arrow at top-right indicates direction of applied magnetic field. In pictures on right, dashed lines and small arrows indicate domain-wall contrast and magnetization, respectively. Arrowheads indicate SiO_2 precipitates. Captured images were taken by video camera on small fluorescent plate of transmission electron microscope. Small black circle in center of image is mark printed on small fluorescent plate. Ring-shaped contrast around mark is artifact caused by electron irradiation on small fluorescent plate. Permission obtained from Elsevier [17]

Fig. 3.50 Video: Dynamic motion of domain walls in doubly oriented electrical steel sheet

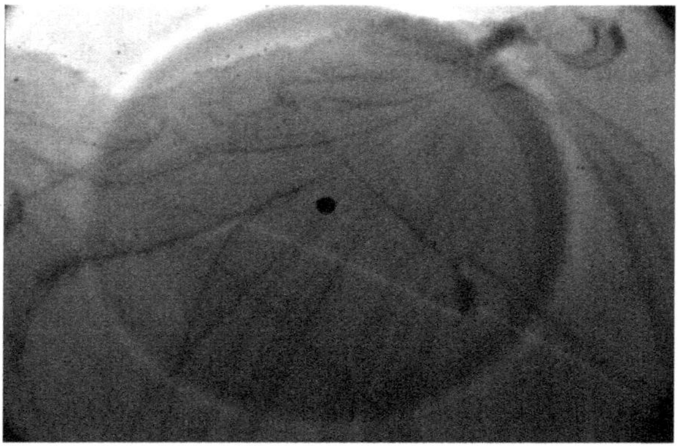

strain field around them. These experimental results can be used to improve not only the mechanical properties, but also the magnetic properties (see a video of the dynamic motion of domain walls in Fig. 3.52).

Magnetization Process of Nd-Fe-B Magnet

To conduct a dynamic investigation of the magnetization process requires the application of a strong magnetic field to the specimen inside a transmission electron microscope. In case of an electromagnet placed in the specimen holder, the maximum magnetic field is limited to about 0.05 T. A unique technique to produce a strong localized magnetic field has been introduced [87], where a small sharp magnetic needle of Nd–Fe–B was utilized to produce the external magnetic field. The needle position can be controlled using piezoelectric elements (i.e., the magnetic field can be increased by moving the needle to the specimen) as shown in Fig. 3.53.

Lorentz microscope images of the nucleation process in $Nd_2Fe_{14}B$ for magnetization reversal in Fig. 3.54 [88] where the images were captured from videotape. The passage of time is indicated in the top-right corner of each image. The speed of the needle was estimated to be about 25 nm per frame, i.e., 0.033 s. An electromagnet was used to apply a magnetic field parallel to the direction indicated by the large blue arrow (H_{ext}) in Fig. 3.54a, after which it was removed.

The magnetization direction in grains G1 and G2 was considered to be parallel to the axis of easy magnetization in each grain. However, the magnetized sharp needle approached the thin-foil specimen in the direction indicated by the large red arrow in Fig. 3.54b. The magnetic field at the top of the needle was estimated to be about 560 kA m^{-1} by simulation. The magnetic field at the specimen position in Fig. 3.54a was about 120 kA m^{-1}. The thin-foil specimen was magnetized and thus, no white lines or black bands corresponding to magnetic domain walls, were observed until Fig. 3.54a. A pair of domain walls abruptly appeared immediately after Fig. 3.54a, as shown in Fig. 3.54b. The domain walls, which appear as white lines and black bands, are shaped as straight lines. It is noted that the nucleation of magnetization reversal occurs

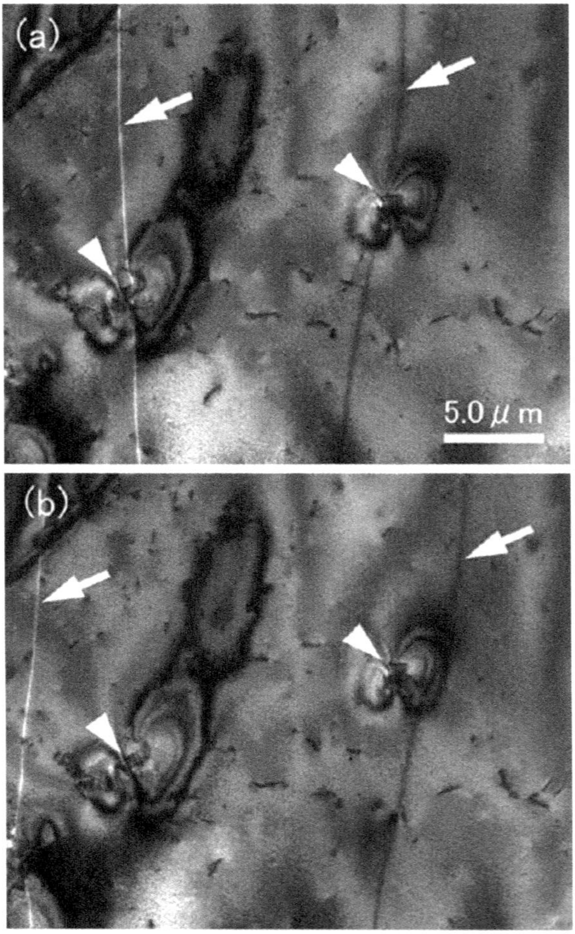

Fig. 3.51 Lorentz micrographs of a non-oriented electrical steel sheet observed under two-beam condition of a bright-field image. Arrows and arrowheads indicate domain-wall contrasts and two AlN particles, respectively. Permission obtained from Elsevier [17]

abruptly with a large volume of paired, reversed domains that are formed from and along the grain boundary. A pair of magnetic domain walls is formed preferentially, not in grain G2 but in grain G1 (Fig. 3.54b). This result indicates that the nucleation of magnetization reversal is strongly dependent on the shapes and properties of the grain boundaries. Thus, control of the shapes and properties of the grain boundaries is important to improve the magnetic properties, e.g., the coercivity of these hard magnetic materials. After the reversed domains are formed, the magnetic domain walls move continuously [88] (see videos of the dynamic motion of domain walls in Fig. 3.55 (first half) and 3.56 (second half)).

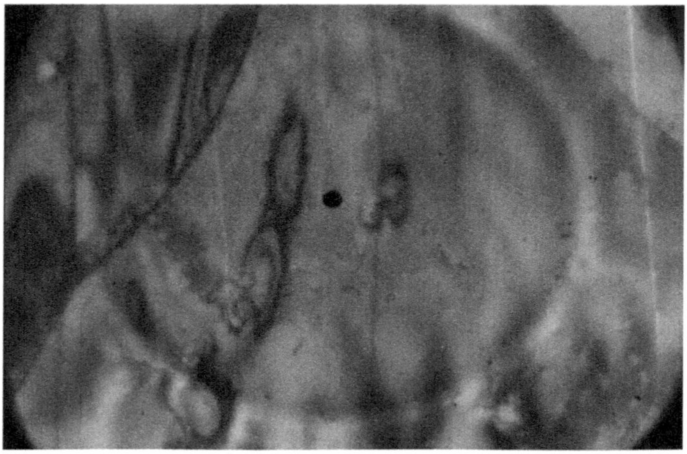

Fig. 3.52 Video: Dynamic motion of domain walls in non-oriented electrical steel sheet

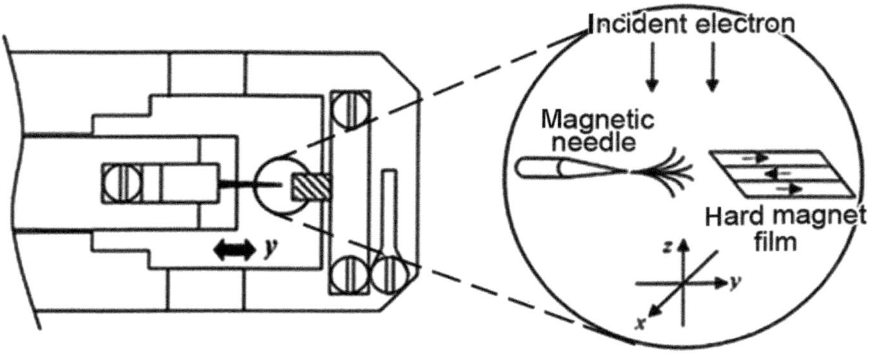

Fig. 3.53 Schematic of part of a specimen holder equipped with a sharp magnetic needle. Permission obtained from Elsevier [17]

3.4 Bulk Materials

3.4.1 Secondary Electron Manipulation with $BaTiO_3$

The cleavage property of $BaTiO_3$ was utilized to investigate the charging effect and motion of secondary electrons. First, an attempt was made to simply reproduce the result obtained for a microfibril with two branches, i.e., the accumulation of secondary electrons between the branches, using an FIB method to control the specimen shape.

A scanning ion microscopy (SIM) image of a $BaTiO_3$ specimen with two branches at the top is shown in Fig. 3.57a. The thickness of the specimen along the incident

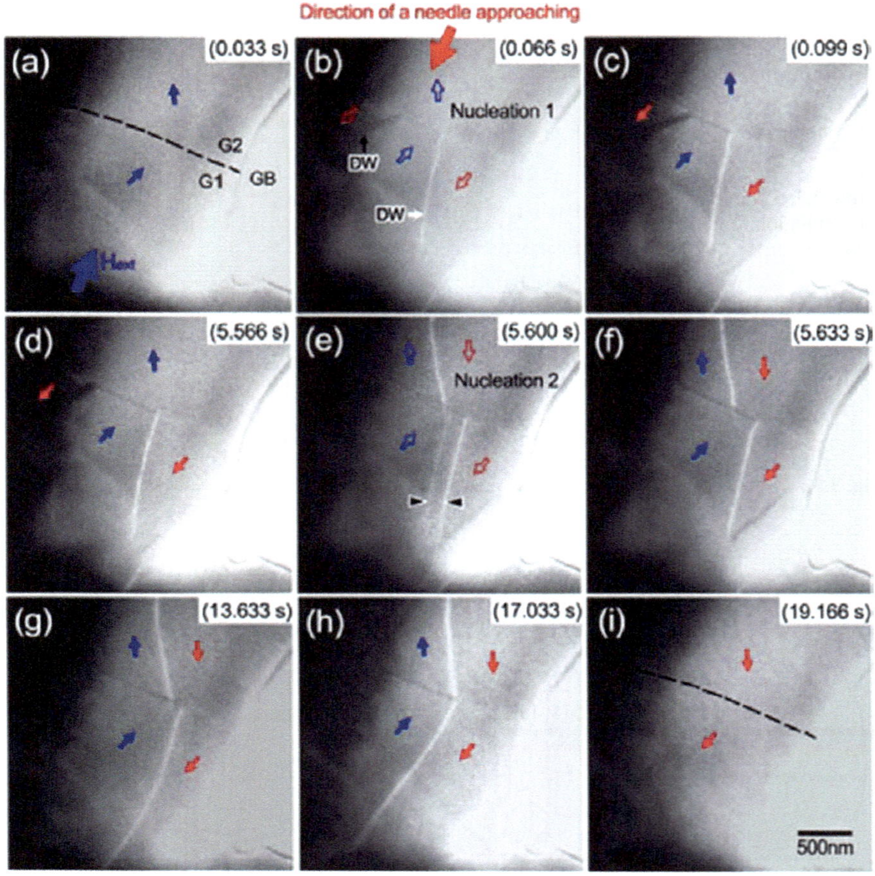

Fig. 3.54 Lorentz microscope images captured from a videotape, showing nucleation process for magnetization reversal in $Nd_2Fe_{14}B$: **a** Remanent state. **b** Reversed domains form in grain G1 from grain boundary. **c**, **d** Magnetic domain walls change gradually. **e** When one of domain walls disappears in grain G1, a domain wall appears in grain G2. **f**, **g** Domain walls in grains G1 and G2 move continuously. **h** Domain walls become connected at grain boundary and **i** move to left until they disappear. Passage of time is indicated in parentheses. GB and DW represent grain boundaries and domain walls, respectively. Blue and red arrows indicate direction of magnetization in each domain. Open arrows indicate magnetization direction at initiation of nucleation. Permission obtained from Elsevier [17]

electron beam direction was about 6 μm. Figure 3.57b shows a reconstructed phase image of the top-edge region in Fig. 3.57a, where the branches are shown in purple. The electric potential of the specimen was estimated to be 2.5 V from simulation in Fig. 3.57c. In Fig. 3.57d, unlike the case of microfibril of sciatic nerve tissue noted above, secondary electron accumulation was not observed between the branches.

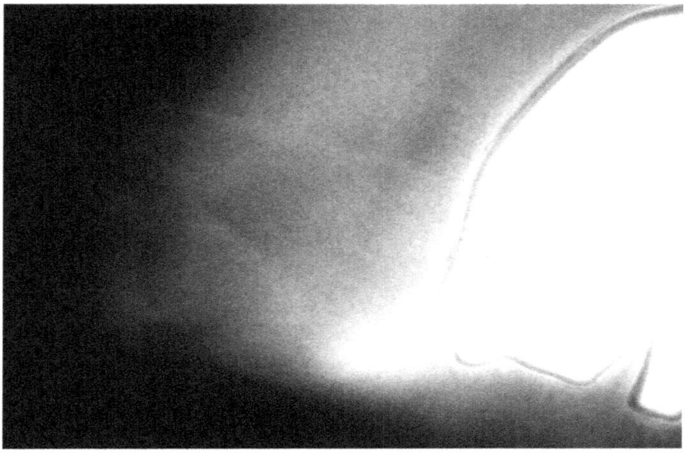

Fig. 3.55 Video (first half): Dynamic motion of domain walls in Nd-Fe-B. The video was obtained in collaboration with Dr. Park

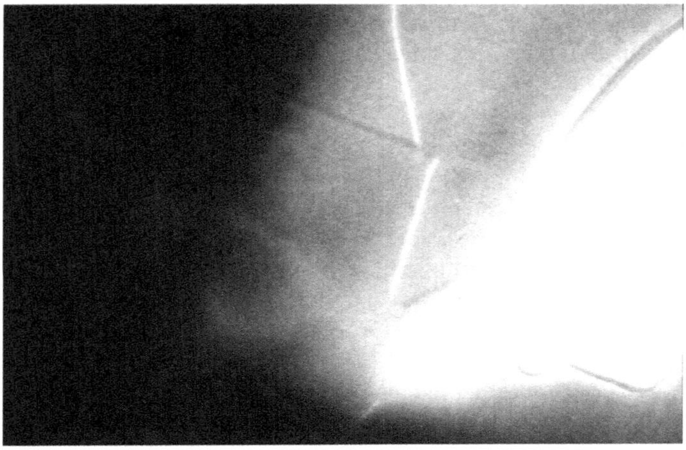

Fig. 3.56 Video (Second half): Dynamic motion of domain walls in Nd-Fe-B. The video was obtained in collaboration with Dr. Park

This difference in the electron distribution is attributed to the formation of a conductive layer on the specimen surface during FIB thinning. After successive electron microscope observation, part of the right branch was broken by mechanical shock during the treatment and transfer of the specimen, as evident in Fig. 3.57e. The electric potential in most of the branch region is about 0.1 V due to substantial carbon contamination. On the other hand, the fresh surface edge formed by the cleavage has a localized contour line that corresponds to an electric potential of 0.28 V, as shown in simulation of Fig. 3.57f. The reconstructed amplitude image in Fig. 3.57g clearly

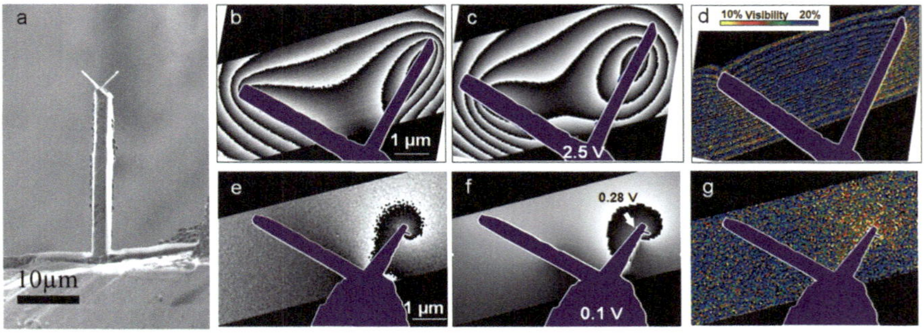

Fig. 3.57 Charging effect and secondary-electron distribution of $BaTiO_3$. **a** SIM image showing branch region of $BaTiO_3$. **b** Reconstructed phase image of $BaTiO_3$. **c** Simulated phase image of same region. Electric potential of surface was assumed to be 2.5 V. **d** Reconstructed amplitude image of $BaTiO_3$. **e** Reconstructed phase image of specimen with cleaved edge in right branch. **f** Simulated phase image of same region. Electric potential of surface was assumed to be 0.1 V, except at cleaved edge, where it was assumed to be 0.28 V. **g** Reconstructed amplitude image of specimen with cleaved edge. Permission obtained from Elsevier [17]

shows bright colored regions around the fresh surface area [89]. These experimental results demonstrate that the collective motion of electrons is sensitive to the surface conditions of insulating materials.

Taking these results into account, the shape of the specimen can be artificially controlled to form a localized fresh surface for clearer visualization of the secondary-electron distribution. Figure 3.58a shows an SIM image of a $BaTiO_3$ rod with a triangular prism shape prepared using an FIB method. The red arrow indicates the thin neck region formed at the wedge tip. The specimen was cleaved immediately after FIB processing by application of a mechanical shock with a glass fiber. The cleaved specimen was then mounted on a specimen holder.

Figure 3.58b shows a hologram of the upper region of the cleaved specimen. The reconstructed phase image in Fig. 3.58c shows that the electric potential reached a high value (*ca.* 13 V) in the cleaved region with the fresh surface from the simulated phase image shown in Fig. 3.58e. The specimen thickness in the cleaved area was set to 200 nm in the simulation, while the thickness of the other regions was set to 2.9 μm, and the electric potential for the surface was set to 0.5 V. In the reconstructed amplitude image shown in Fig. 3.58d, bright red and yellow speckles are evident on the cleaved surface and extend over a wide area. The variations in visibility in the hologram, measured along paths B–A and C–D (see Fig. 3.58b) are compared in Fig. 3.58f. The visibility is minimal in region C, i.e., near the freshly cleaved surface, and increases toward region D while remaining lower than that between B and A [89].

The results shown in Fig. 3.58 indicate that a secondary-electron manipulator can be developed by utilizing the phenomena of secondary electron accumulation being sensitive to the presence of metallic elements on the surface of charged insulating materials. An insulating $BaTiO_3$ rod specimen such as that shown in the SIM image

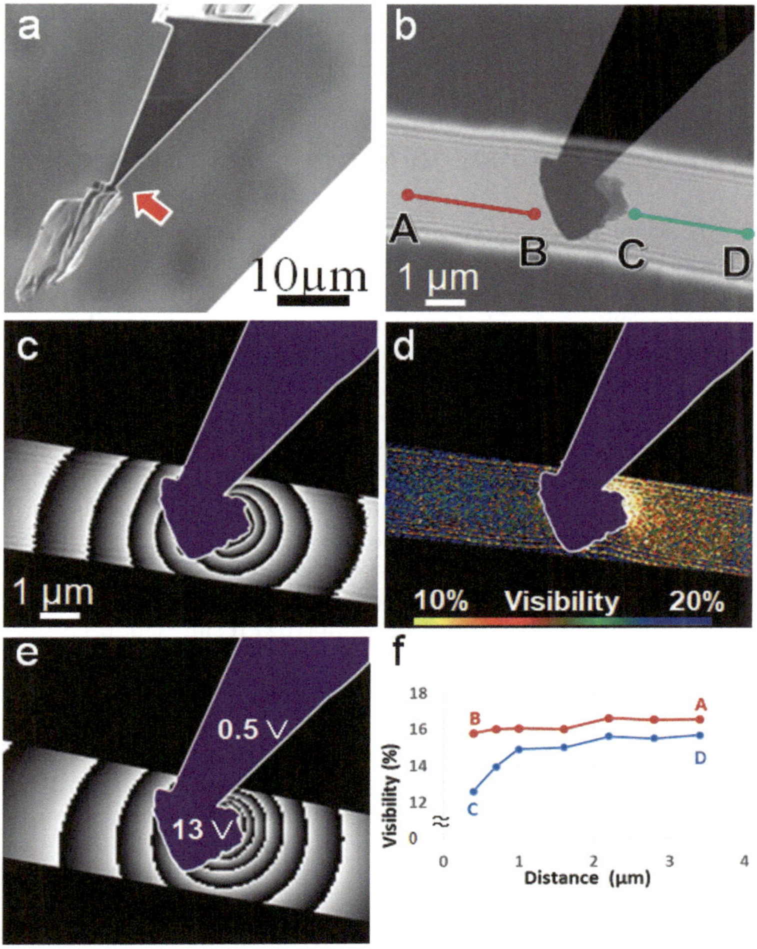

Fig. 3.58 Distribution of secondary electrons around cleaved $BaTiO_3$ rod tip. **a** SIM image of $BaTiO_3$ rod before cleavage. **b** Hologram of $BaTiO_3$ rod tip after cleavage. **c** Reconstructed phase image of $BaTiO_3$ rod tip obtained from hologram. **d** Reconstructed amplitude image obtained from hologram. **e** Simulated phase image corresponding to (**c**). **f** Visibility as function of distance from specimen between points indicated in (**b**). Permission obtained from Elsevier [17]

in Fig. 3.58a was prepared using a focused Ga-ion beam. A mechanical shock was applied to the rod (triangular prism shape) to cleave the specimen in the thin neck region at the wedge tip, which resulted in the formation of a localized clean surface. Observation of this specimen by electron holography revealed that secondary electrons accumulated locally around the clean surface region, as shown in Fig. 3.58d. Mounting this specimen in a specimen holder with a piezodriving probe (Fig. 3.53) and moving this specimen to semiconductor specimens enable direct observation of the interactions of the secondary electrons with the specimen surfaces and various internal defects.

3.4.2 Control of Electron Distribution on Mica Surface

As noted, the accumulation and collective motions of electrons around various positively charged insulators have been observed experimentally through the amplitude reconstruction process in electron holography. These experiments were performed using electron microscopy with thin insulating surface specimens. In this section, similar effects of secondary electrons, which are attracted to the bulk surfaces of insulators, can be observed by SIM with an FIB system, as shown in Fig. 3.59 [90]. The common feature of the phenomena observed using electron microscopy and SIM is the positive charging effect of insulators. A Ga^+-ion probe is used to scan the bulk insulator surface during SIM observation. Secondary electrons are emitted from the bulk surface at each point in the scanned area, while Ga^+ ions accumulate in the bulk specimen, which results in positive charging.

An FIB system was used to deposit two W islands on a mica substrate in the form of the letter "e" and a superscript minus symbol ("–"). SIM observation was then performed for 48 s with a nominal Ga^+dose of 0.027 (C m^{-2}), or 0.17 (ion nm^{-2}). Figure 3.60a shows the initial state. When the "–" island was touched with a 5 V biased probe, it became brighter, as did the "e" island (Fig. 3.60b), which indicated charge transfer between the "–" island and the "e" island along with the presence of Ga and W elements on the mica substrate. This matches the results obtained for the epoxy thin film (Fig. 3.45). The presence of W was due to its spread during the deposition process. When the "e" island was touched by the probe, it became much brighter, and the "–" island became slightly brighter (Fig. 3.60c). Finally, when the "–" island was again touched by the probe, it became brighter; however, the "e"

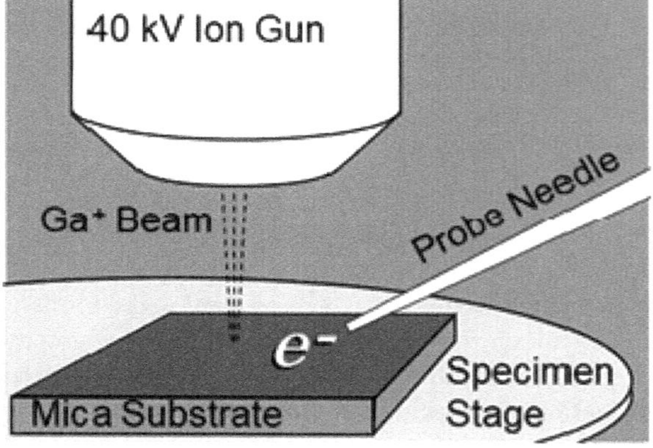

Fig. 3.59 Experimental setup in which "e^-" pattern was drawn on insulating mica surface using W deposition and probe needle

island did not (Fig. 3.60d). The different result shown in Fig. 3.60d compared with those in Fig. 3.60b, c is attributed to enhanced sputtering of the specimen surface by Ga^+ ions [90]. Figure 3.61 shows a video of SIM observation.

The transfer of charge between the islands is considered to correspond to the so-called tunneling phenomenon. This phenomenon largely affected by the existence of the metallic elements is discussed with the SLW (Spinning Linear Wave) model in Sect. 5.2.3.

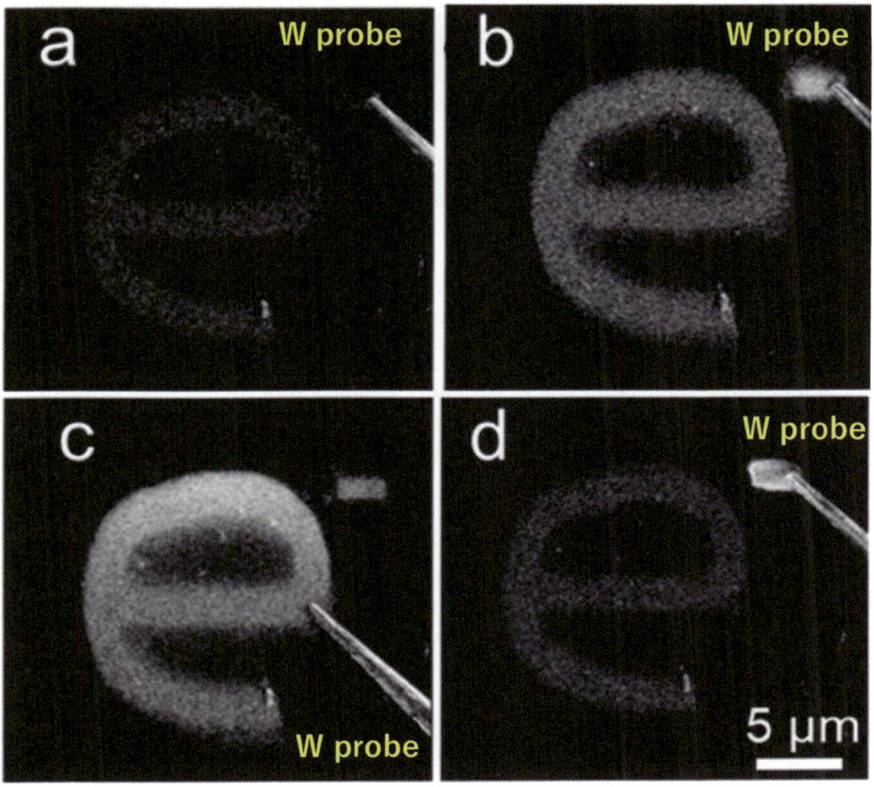

Fig. 3.60 Successively recorded SIM images of "e^-" characters prepared on mica substrate by W deposition. **a** W islands not touched by +5 V-biased probe needle. **b** "−" character touched by probe. **c** "e" character touched by the probe. **d** "−" character touched by the probe with a 0.17 ion/nm^2 additional Ga^+ dose after scene (**b**)

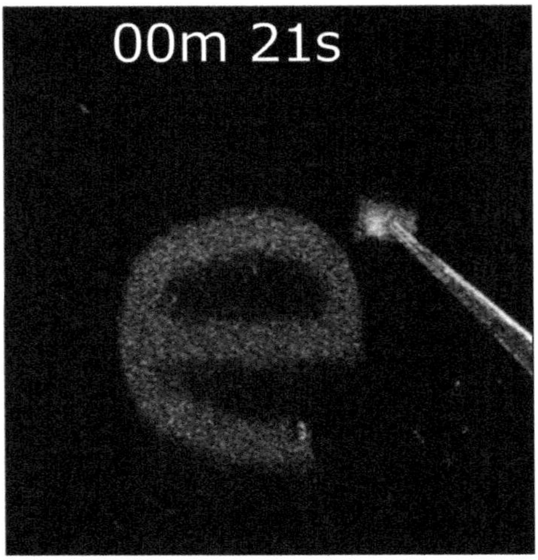

Fig. 3.61 Video: Contrast change in SIM image of e^-. The video was obtained in collaboration with Ms. Shimada

3.4.3 Meissner Effect and Quantized Magnetic Flux Distribution Around Y-Ba-Cu–O

The temperature-induced magnetic structure changes in a high-temperature superconductor can be investigated using electron holography. The critical current density J_c, in type-II superconductors is the most important property related to their practical use because a high J_c can be achieved through pinning interaction between the quantized magnetic flux (fluxons) that infiltrates the superconductor and the pinning centers [91, 92]. Improvement in the properties of superconductors requires understanding of the magnetic flux behavior. Therefore, electron holography was combined with SIM to visualize the magnetic flux distributions in the high-temperature superconductor Y–Ba–Cu–O with non-superconducting particles [93].

The bulk material was fabricated via a quench-and-melt growth process and consisted of a superconductive $YBa_2Cu_3O_{7-y}$ matrix (123 phase) and non-superconductive Y_2BaCuO_5 particles (211 phase) [94, 95]. A small square-column specimen (2 × 2 × 25 μm) was cut from the bulk material using an FIB with the *c*-axis perpendicular to the long axis of the square-column shape. The specimen contained one or two 211-phase particles in the direction of the incident electron beam.

The lower-left diagram in Fig. 3.62 shows a schematic of the setting for specimen analysis. The specimen was fixed on the side edge of a semicircular Cu grid by tungsten deposition using an FIB. The *c*-axis of the 123 phase was aligned horizontally and an external magnetic field was applied along the *c*-axis. Electron holography was performed using a 300 kV transmission electron microscope equipped with a liquid-He cryogenic system with magnetic coils to apply an external magnetic field [96]. The specimen holder was inserted above the pole pieces of the objective lens to reduce the magnetic field at the specimen position and thus preserve the magnetic

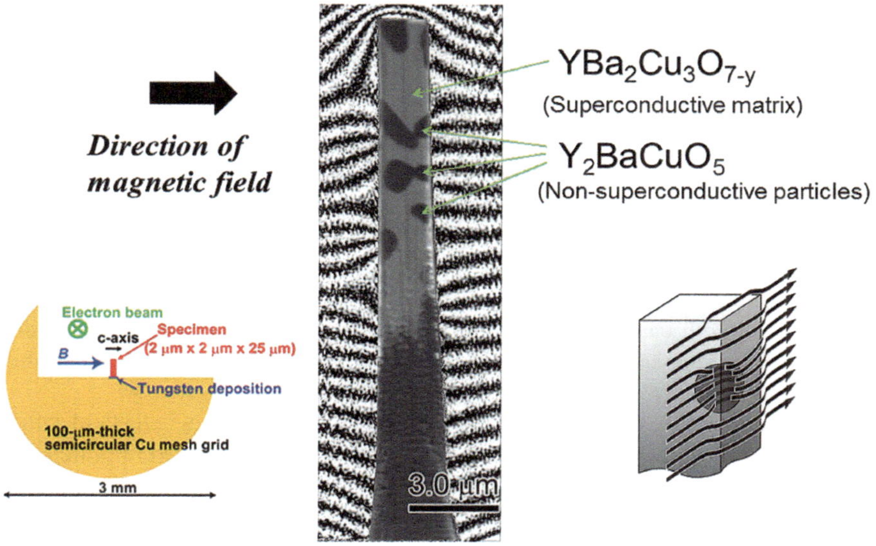

Fig. 3.62 (left) Schematic of experimental setting. Specimen was fixed on side edge of semicircular Cu mesh grid, and *c*-axis of superconductive phase was set parallel to external magnetic field. (center) Reconstructed phase image of square-column-shaped Y–Ba–Cu–O specimen in external magnetic field of 0.01 T at 12 K. Phase was amplified by a factor of 4. Phase image is overlain with SIM image of specimen to enable comparison between magnetic flux distributions and 211-phase particles. (right) Schematic of magnetic flux lines around specimen with 211-phase particle located at specimen edge. Permission obtained from Elsevier [17]

structure in the specimen. During reconstruction of the phase images, the phase information above the critical temperature was subtracted to remove phase modulation due to the charging effects of the insulating layers on the surface of the specimen; these layers were introduced during the FIB thinning process.

Electron holography observations were performed in a magnetic field of 0.01 T. When the phase reconstruction process was performed under a constant magnetic flux applied to both the object wave and the reference wave, the reconstructed phase image showed a deviation of the magnetic flux from the zero field. The addition of a constant magnetic flux in the specimen plane with a thickness corresponding to the particle diameter was useful to visualize the magnetic flux distribution around the particle. The phase of the reconstructed phase image was amplified by a factor of four. The center of Fig. 3.62 shows a reconstructed phase image of a square-column-shaped Y–Ba–Cu–O specimen. The phase image was overlapped with the SIM image of the 211-phase particles to facilitate comparison. The heavier components in the SIM images are darker [97]; therefore, the 211 phase has darker contrast than the 123 phase, which makes identification of the 211-phase particles easier. The lines of magnetic flux detoured around the top-edge region of the specimen due to the Meissner effect and they were denser at the position of the 211-phase particles, as clarified from the side-view SIM images of the specimen (Fig. 3.64a, c) [93]. The lower-right diagram in Fig. 3.62 shows a schematic of the magnetic flux lines around a specimen with a 211-phase particle located at the specimen edge. Figure 3.63a shows the data observed at 12, 22, 35, 44, 55, 67, 76, and 87 K. Magnified views of the dashed rectangular regions in Fig. 3.63a, in which the density of magnetic flux lines is higher on the 211-phase particle due to the Meissner effect, are shown in Fig. 3.63b. The phase images shown in Fig. 3.63b were amplified by a factor of 8. Therefore, the phase shift between two black lines is $\pi/4$. The phase shift in the blue region, which contains three black lines, is $\pi/2$. The distance between the red arrowheads in Fig. 3.63b increases from 55 to 87 K; this distance corresponds to a phase of $5\pi/8$ rad in the vicinity of a 211-phase particle. This temperature dependence of the contour-line width can also be explained by the concept of magnetic penetration depth. At high temperatures, the magnetic penetration depth is increased and thus the distribution of the magnetic field becomes more uniform.

The remanent state was examined by removal of the external magnetic field, which led to the results shown in Fig. 3.64. The reconstructed phase image (amplified by a factor of two) in Fig. 3.64b shows the distribution of magnetic flux lines outside the specimen at 13 K; this image overlaps the SIM image of the front of the specimen. The reference wave was not affected by the external magnetic field, because the field was turned off; therefore, the contour lines can be considered as a projection of the magnetic flux lines. The phase was amplified by a factor of two; therefore, the magnitude of one fluxon ($h/2e$) corresponds to one period of the contour lines

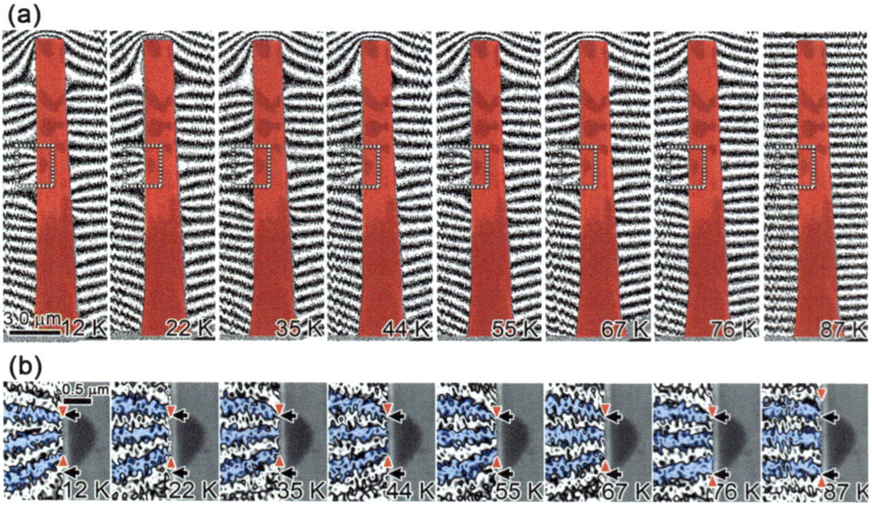

Fig. 3.63 **a** Reconstructed phase images of the specimen in an external magnetic field of 0.01 T. The phase was amplified by a factor of 4. Temperature is indicated at lower right. **b** Magnified view of dashed rectangular regions in (**a**). Phase was amplified by a factor of 8. Black arrows indicate edges of 211-phase particle. Permission obtained from Journal of American Institute of Physics [93]

[41, 93, 98]. Figure 3.64a, c show SIM images of the left- and right-side views, respectively, where the fluxons are stably pinned exactly at the 211-phase particle sites. This pinning occurred because these particles in the middle of the specimen were surrounded by the superconducting 123 phase. As demonstrated in this study, the combination of electron holography to characterize the magnetic flux distribution and SIM imaging to characterize the microstructure is useful to clarify the pinning positions of magnetic flux. This technique can be extended to analyze the interactions between the magnetic flux and various lattice defects to clarify and improve the J_c of high-temperature superconductors.

Figure 3.65a shows the data obtained at 13, 28, 45, 59, 74, and 88 K. Figure 3.65b shows enlarged reconstructed phase images of the dashed rectangular regions in Fig. 3.65a. The phases were amplified by a factor of 8, and the regions that correspond to the black contour lines in Fig. 3.65a are shown in blue. The magnetic flux distribution changed gradually up to 59 K. Around 74 K, it began to change significantly, and the magnetic flux disappeared at 88 K. The distribution of the fluxon also widened around 74 K. It is considered that the magnetic penetration depth had become large due to the increase in temperature. It is noted that the magnetic penetration depth of the 123 phase increases rapidly in the vicinity of 60 K. The increase

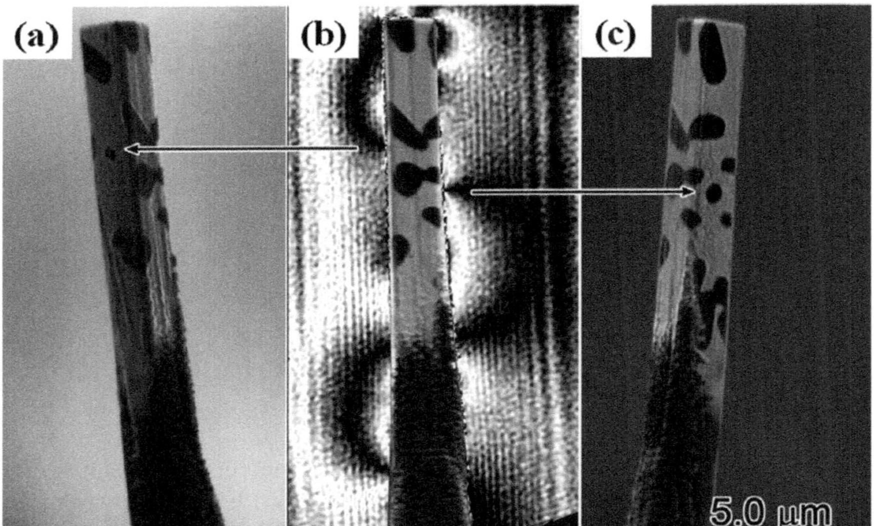

Fig. 3.64 **a** SIM and reconstructed phase images of square-column-shaped Y–Ba–Cu–O superconductor in the remanent state. In center image (**b**), reconstructed phase image at 13 K overlaps SIM image of specimen for comparison between distributions of magnetic flux and 211phase particles. SIM images (**a**) and (**c**) are left- and right-side views of specimen. Both images were obtained with tilting angles of ±52° from center image. Emission points of fluxon in (**b**) correspond to insulating particles indicated by arrows in (**a**) and (**c**). Permission obtained from Journal of American Institute of Physics [93]

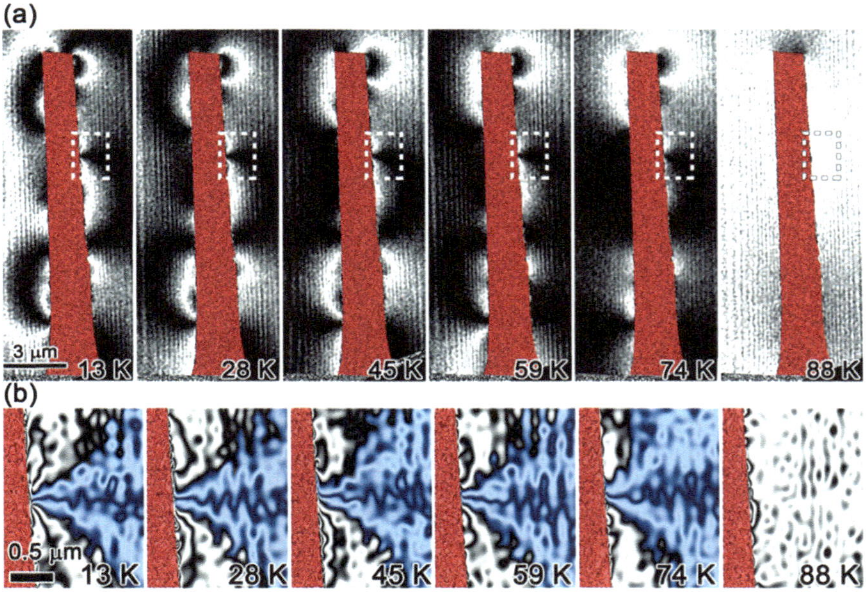

Fig. 3.65 **a** Reconstructed phase images of specimen in remanent state. Phase was amplified by a factor of 2. Temperature is indicated at lower right. **b** Magnified view of dashed rectangular regions in (**a**). Phase was amplified by a factor of 8. Permission obtained from Journal of American Institute of Physics [93]

of the blue area from 59 to 74 K in the reconstructed phase image in Fig. 3.65b corresponds to this tendency.

3.5 Devices

3.5.1 *Electric Field*

A double probe piezodriving holder (see Fig. 4.11) was modified to introduce a laser irradiation port in its left arm (arm 1), as shown in Fig. 3.66 [99]. A light convergence function can be added at the top of the laser irradiation port. A plastic fiber is placed inside arm 1 to introduce the laser beam to the holder. Various laser systems with different energies can be placed outside the microscope. An enlarged image of the top portion of the specimen holder is shown on the right-hand side of Fig. 3.66. The laser irradiation port and the probe can be manipulated independently in three dimensions by controlling the motions of arms 1 and 2. These two arms are driven by micrometers and piezoelectric elements that are located in the tail portion of the holder. Three manual micrometers are used to effect coarse movement of the laser

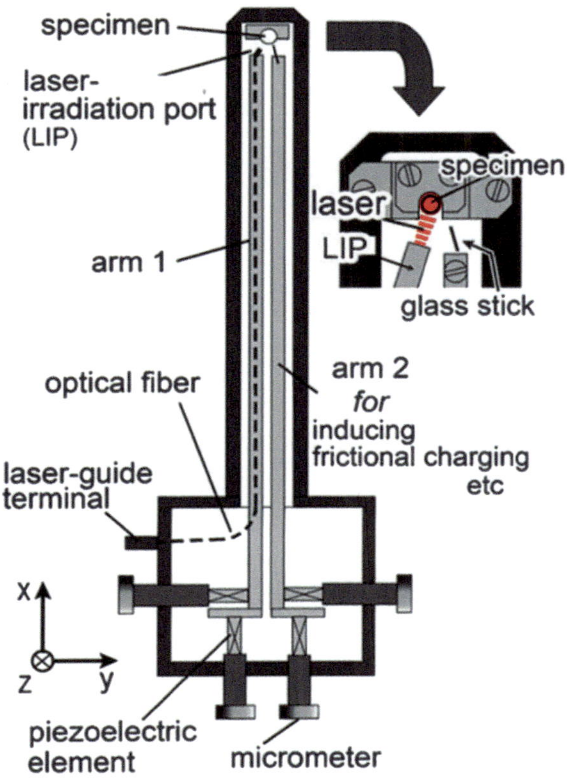

Fig. 3.66 Schematic illustration of modified specimen holder equipped with a piezodriving probe and a laser irradiation port. On righthand side, enlarged diagram of top portion of specimen holder is shown. Permission obtained from The Japanese Society of Microscopy [99]

irradiation port and the probe, while three piezoelectric elements are used to effect fine movement. Figure 3.67a shows an image of the top portion of the specimen holder. The specimen appears red when it is irradiated by a red laser beam, as shown in Fig. 3.67b.

The specimen holder was used to investigate the discharge effect in commercial organic photoconductors, which are widely utilized in electrophotography. It should be noted that the discharge effect in organic photoconductors has not been directly analyzed to date. Electrostatic latent charges in most laser printing systems are produced through the so-called corona discharge effect on the surface of an organic photoconductor, which is comprised of a charge-transport layer, a charge-generation layer, an undercoat layer and a substrate. Instead of using the corona discharge effect to generate the electrostatic latent charges, frictional charging was employed along with a small glass rod, as shown in Fig. 3.68a. An electric shield was placed above the specimen in a manner similar to that in an experiment on toner particles in Sec. 4.2.1 [100] to avoid charging caused by the emission of electron-induced secondary electrons. A Mo plate was placed in one of the piezodriven arms to shield the toner particles. A Mo plate was also directly fixed at the top of the organic

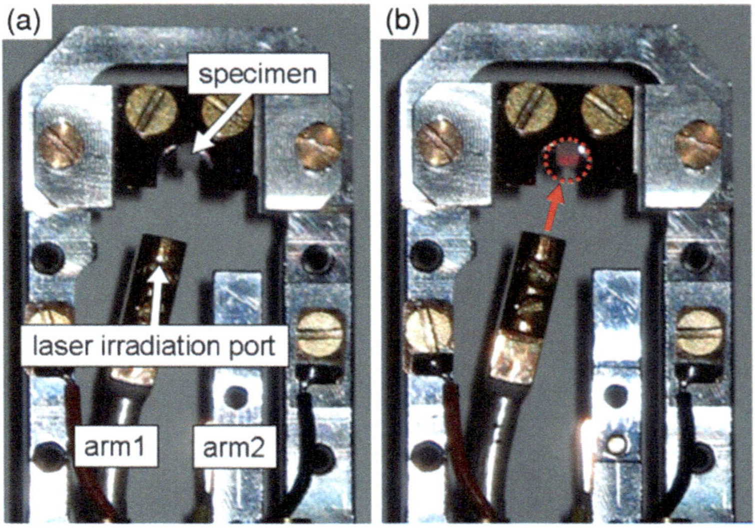

Fig. 3.67 Images of top portion of specimen holder with laser beam **a** off and **b** on. In **b**, red arrow indicates direction of laser beam. Permission obtained from The Japanese Society of Microscopy [99]

photoconductor, as shown in Fig. 3.68a, which is referred to as 'shield' in the figure. After frictional charging using a glass rod, negative charges were expected to appear on the surface of the organic photoconductor, as shown in Fig. 3.68b. Electron holograms were observed, and reconstructed phase images were obtained, as shown in Fig. 3.69a, to investigate the formation of negative charges. The reference wave modulation caused by the electric field was neglected in the reconstruction process. Figure 3.69b shows a comparison of the electric potential distributions projected onto the incident electron beam along the line *x*–*y* for two instants of time, 0 and 3 min. It should be noted that the electric potential produced by frictional charging did not change significantly for periods of time < 3 min under the present experimental conditions. On the other hand, when the laser light was radiated onto the surface of the organic photoconductor, the electric potential around the organic photoconductor was observed to change considerably, as shown in Fig. 3.70a, b. It is likely that the electric potential distribution in Fig. 3.70a is not completely identical to that in Fig. 3.69a, although both of the observations represent the state after the frictional charging. The deviation is presumably due to differences in the experimental conditions between these observations, e.g. such as the position of the photoconductor with reference to the edge of the Mo shield. Figure 3.71a shows the change in the projected electric potential distribution before and after laser irradiation. The irradiation time for the laser beam was 30 s, and the total time elapsed during the experiment, including the time of laser irradiation and the time taken for hologram observation, was within 2 min. The change in the discharge effect with time was negligible according to the data in Fig. 3.69. Figure 3.71a shows that the electric

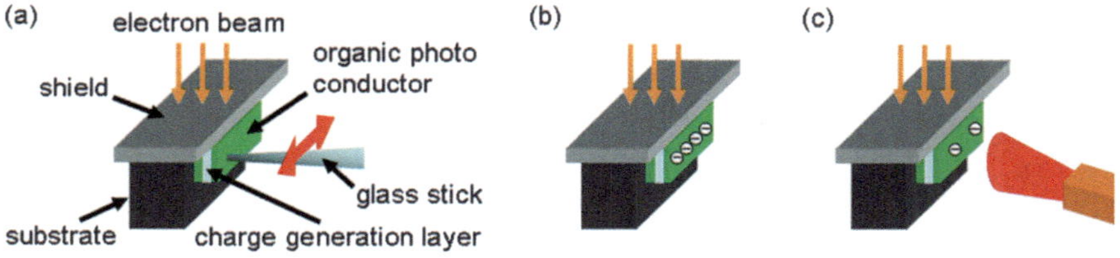

Fig. 3.68 **a** Schematic illustration of frictional charging by using a small glass rod. **b** Because of frictional charging, negative charges appear on surface of organic photoconductor. **c** Number of negative charges is reduced by the generation of electron–hole pairs in charge-generation layer of organic photoconductor. Permission obtained from The Japanese Society of Microscopy [99]

field caused by the negative charges around the organic photoconductor is reduced due to laser irradiation (Fig. 3.68c). Some of the negative charges may have disappeared because of electron–hole annihilation, which involves the generation of holes in the charge-generation layer of the organic photoconductor as a result of laser irradiation (Fig. 3.71b). Therefore, it has been demonstrated that the modified specimen holder with a laser irradiation port is suitable for the investigation of photo-induced phenomena, such as the discharge effect of laser irradiation on organic photoconductors. The reference wave modulation caused by the electric field was taken into account to perform further quantitative analysis. The calculated potential distribution near the point charge is shown in Fig. 3.72a. Note that the phase information is presented in terms of $\cos(\phi)$ in Fig. 3.72, although the results in Figs. 3.69 and 3.70 are given in terms of ϕ. It appears that the lower part in the simulation (Fig. 3.72a) is in agreement with the observation (Fig. 3.72b) when the amount of point charge is assumed to be 3.3×10^{-17} C.

3.5.2 *Magnetic Field*

Figure 3.73a shows the magnetic recording head set on an electron microscope specimen holder and connected to an electric source to visualize the distribution of the magnetic field [101]. The magnetic recording head coated with gold to reduce the electric charging effect was fixed on a side edge of a Si wafer piece with carbon tape. The Si wafer piece was set on an arm of a single-probe piezo-driving specimen holder. The optical microscopy image in Fig. 3.73b shows the appearance of the magnetic recording head. Figure 3.73c shows an enlarged SEM image of the region indicated by the circle in Fig. 3.73b. Tungsten was deposited by the FIB method to mark the position of the magnetic pole.

Figure 3.74a shows the hologram observed around the magnetic pole region without the application of an electric current. The position of the pole is indicated in the lower part of Fig. 3.74a. The region with decreased brightness is attributed to the presence of dust on the magnetic recording head and is indicated by an arrow.

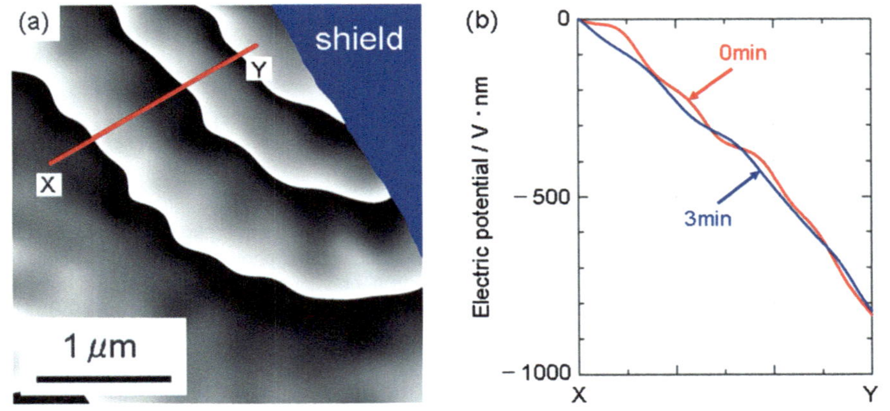

Fig. 3.69 **a** Reconstructed phase image showing electric potential distribution after frictional charging. The phase is amplified by 3. **b** Electric potential distribution projected onto incident electron beam along line *x–y* for 0 and 3 min after frictional charging. In (**b**), for simplicity, electric potential at point 'x' is set '0', since we are particularly interested in change that is observed along the line *x–y*. Permission obtained from The Japanese Society of Microscopy [99]

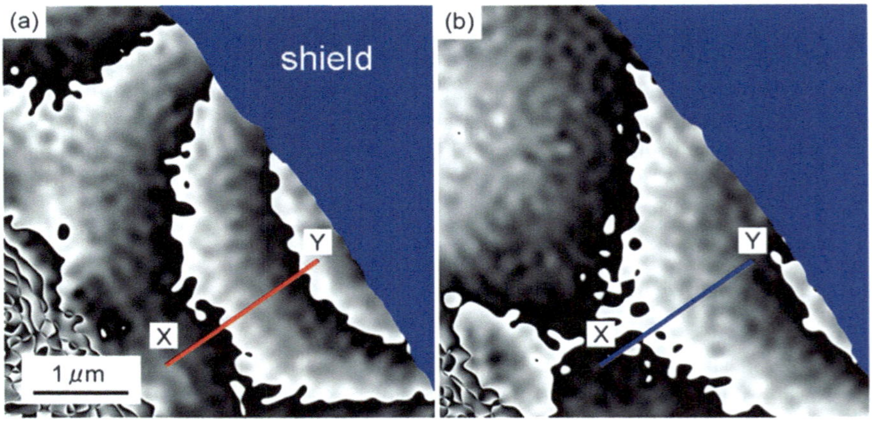

Fig. 3.70 Reconstructed phase images showing electric potential distribution around organic photoconductor **a** after frictional charging (before laser irradiation) and **b** after laser irradiation. In (**a**) and (**b**), phase is amplified by 3. Permission obtained from The Japanese Society of Microscopy [99]

The dust was considered to be formed during the gold coating. Figure 3.74b shows the reconstructed phase image obtained from Fig. 3.74a. The phase shift is wrapped by a modulo operation in which the divisor is 2π, so that (1) the direction of the contour lines corresponds to that of the lines of magnetic flux, and (2) the direction of the magnetic field can be determined from the gradation of contrast between the

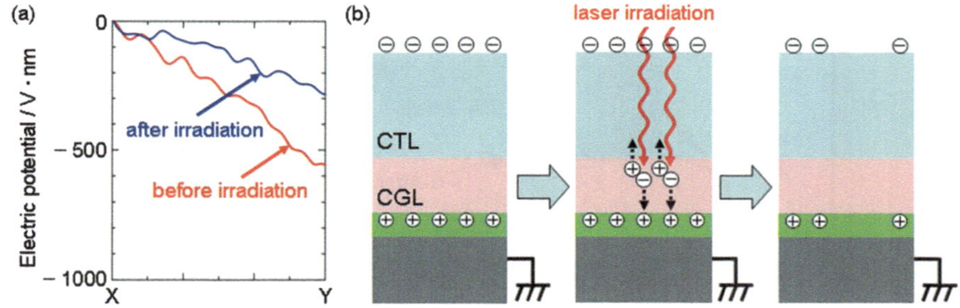

Fig. 3.71 **a** Change in electric potential distribution, which is projected onto incident electron beam, because of laser irradiation. Red and blue curves represent electric potential that is observed along red and blue lines in Fig. 3.70 (namely, line *x*–*y*), respectively. Electric potential at point 'x' is set '0', since we are particularly interested in change that is observed along line *x*–*y*. **b** Schematic illustration of mechanism of negative charge reduction on organic photoconductors upon laser irradiation. CTL: charge-transport layer; CGL: charge-generation layer. Permission obtained from The Japanese Society of Microscopy [99]

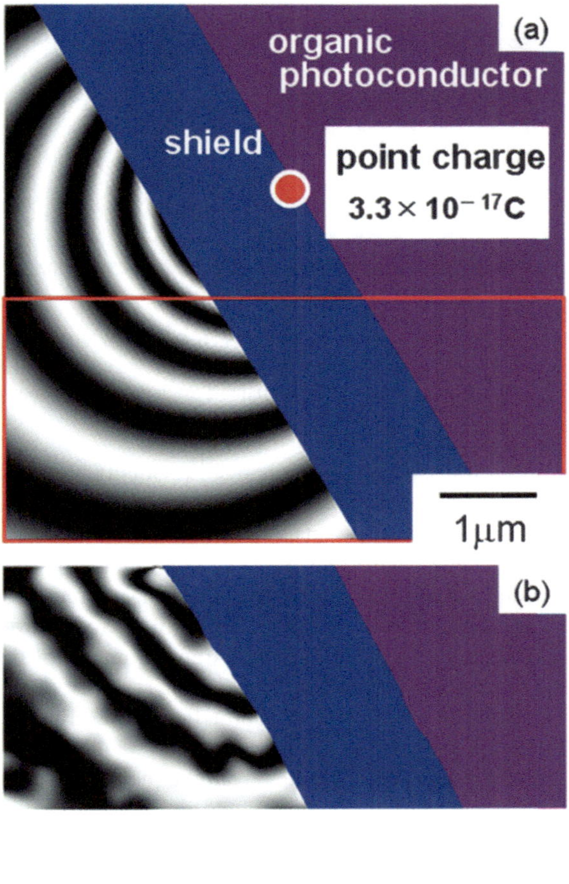

Fig. 3.72 **a** Calculated potential distribution near a point charge that is placed on surface of photoconductor. Effect of reference wave modulation is incorporated. Phase is amplified by 3, and phase information is presented in terms of cos ϕ. **b** Observed potential distribution, which represents state after frictional charging. Phase is amplified by 3. See text for details. Permission obtained from The Japanese Society of Microscopy [99]

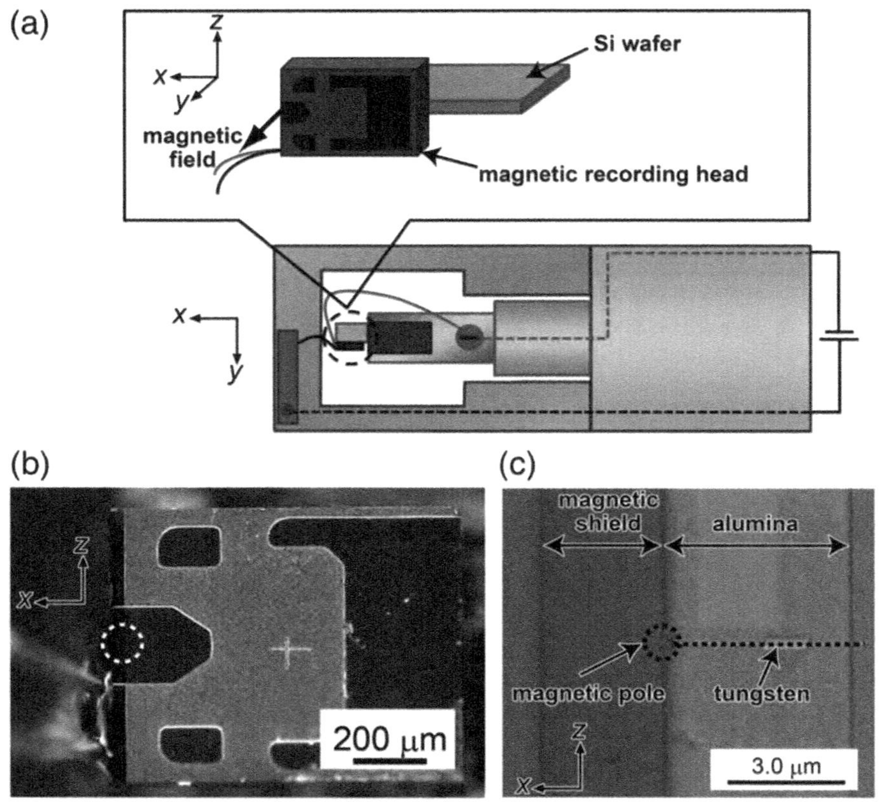

Fig. 3.73 **a** Schematic illustration showing where the magnetic recording head was placed on electron microscope specimen holder. **b** Optical microscope image of magnetic recording head with magnetic pole indicated by a dotted circle. **c** Enlarged SEM image around magnetic pole. Permission obtained from The Japanese Society of Microscopy [101]

contour lines. Some phase change is evident in Fig. 3.74, even though no current was applied to the magnetic recording head, and this contrast is considered to be due to the charging effect. Thus, an attempt was made to cancel the effect of charging from the reconstructed phase images by subtracting the phase information obtained when no current was applied. A reconstructed phase image obtained with an electric current of 1.0 mA is presented in Fig. 3.74c. In the reconstruction process of Fig. 3.74c, the sideband position in the Fourier transformation of the hologram was selected to be the same position as that of Fig. 3.74b. The corrected phase information presented in Fig. 3.74d was obtained by subtraction of the phase information shown in Fig. 3.74b from that in Fig. 3.74c. Therefore, Fig. 3.74d indicates that the relative increase of the magnetic field originates from the magnetic pole. Figure 3.75 shows the experimental results for the reconstructed phase images at various applied currents that were corrected by subtraction of the phase information of Fig. 3.74b. These reconstructed phase images indicate that the magnetic field becomes stronger with an increase in the applied current. The gradation of the contour lines indicates that the direction of the magnetic field reversed when the applied current was reversed. It is noted that the reverse direction of the magnetic field with the same magnitude for the application of the electric currents in Fig. 3.75 indicates that the charging effect in Fig. 3.74b does not change with time or the applied current. The asymmetry in the density of the magnetic flux between the right- and left-hand sides in the vacuum region above the magnetic pole is discussed in the subsequent text with respect to the simulation.

For interpreting the experimental results, an FEM analysis was performed to simulate the reconstructed phase images. The commercial software JMAG for electromagnetic analysis was utilized for this simulation [102]. In the simulation, a simple system that consisted of a permanent magnet and a soft magnetic shield was assumed. The z-component of the vector potential in a three-dimensional calculation region was eventually computed, and the result was then integrated along the z-direction. Contour lines of the integrated z-component of the vector potential [41, 103] were

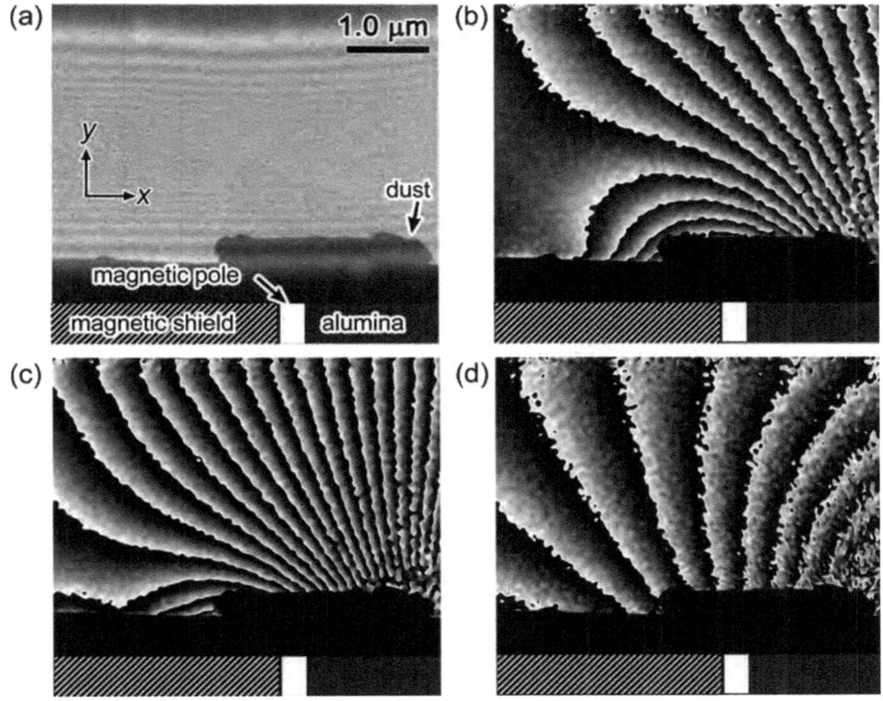

Fig. 3.74 **a** Hologram observed around magnetic pole region when no electric current is applied. Positions of pole, soft magnetic shield, and alumina parts are indicated in the lower part of the figure. **b** Reconstructed phase image obtained from (**a**). **c** Reconstructed phase image obtained with applied electric current of 1.0 mA. **d** Phase image obtained by subtracting phase information in (**b**) from that in (**c**). Permission obtained from The Japanese Society of Microscopy [101]

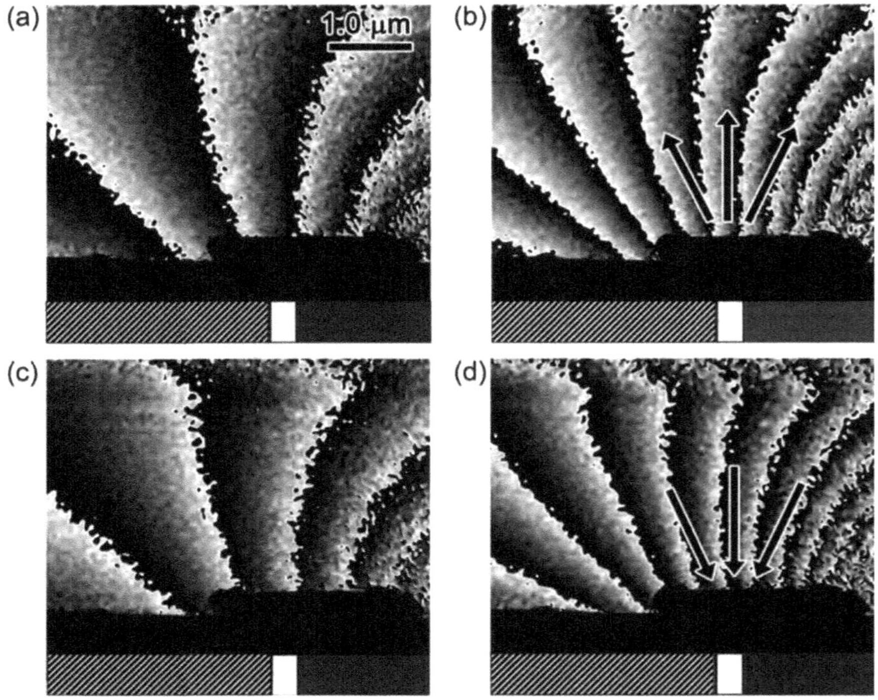

Fig. 3.75 Reconstructed phase images obtained with application of various electric currents: **a** 0.5, **b** 1.0, **c** −0.5, and **d** −1.0 mA. Permission obtained from The Japanese Society of Microscopy [101]

finally obtained, as shown in Fig. 3.76a. For a comparison of the calculated and observed results, the modulation of the reference wave due to the stray magnetic field must be accounted for. Therefore, the vector potential in the reference wave region was subtracted from that in the object wave region shown in Fig. 3.76b [104]. This simulation reveals that the asymmetry of the magnetic field distribution results from the absorption of the magnetic field by the soft magnetic shield located beside the magnetic pole. Such asymmetry of the magnetic field distribution is in good agreement with that evident in the experimental results.

For discussing the experimental results more quantitatively, the reconstructed phase images recorded with various currents applied to the magnetic head were compared with the results of a preliminarily performed FEM simulation while accounting for the detailed model with electric current in the coil and the shapes of the magnetic pole and shield. Figure 3.77a shows an experimental, reconstructed phase image obtained for an applied current of 2.0 mA. It should be noted that the phase shift is wrapped by the cosine function. Figure 3.77b shows the profile for the simulated magnetic flux density along the line indicated by the white arrow in Fig. 3.77a. The asymmetry of the profile, for example, a shoulder on the left side of the peak, originates from the magnetic shield.

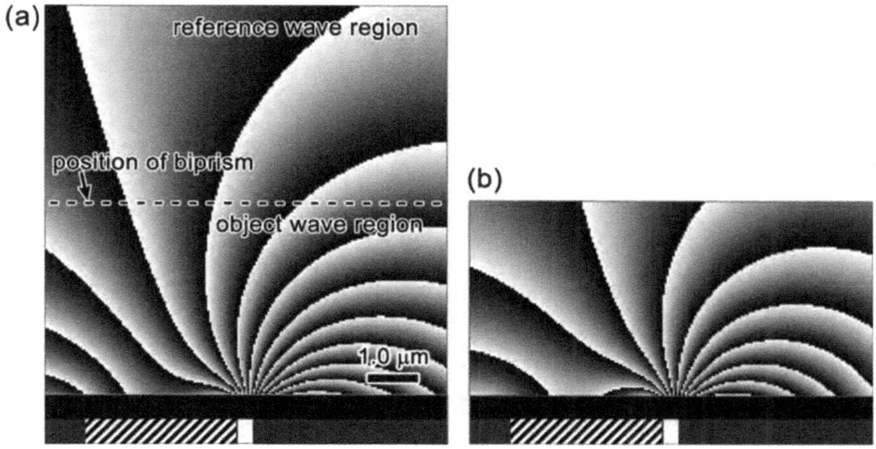

Fig. 3.76 **a** Simulated phase image around magnetic pole. Dotted line indicates position of electron biprism wire. Regions above and below dotted line are reference wave region and object wave region, respectively. **b** Simulated reconstructed phase image where phase information of reference wave region was subtracted from that of object wave region. Permission obtained from The Japanese Society of Microscopy [101]

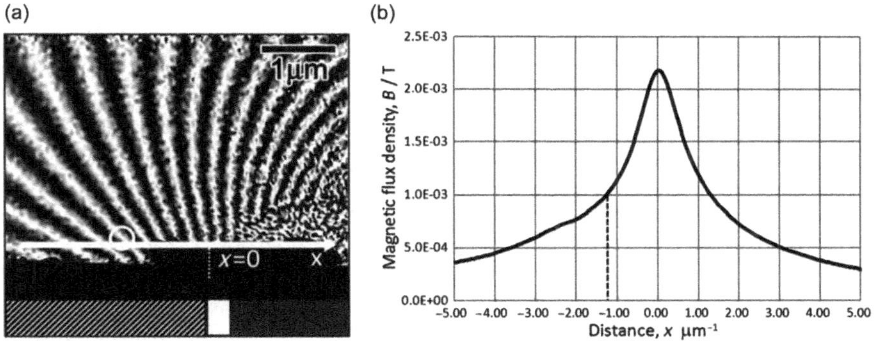

Fig. 3.77 **a** Reconstructed phase image obtained with applied electric current of 2.0 mA. **b** Simulated magnetic flux density profile along white arrow in (**a**). Permission obtained from The Japanese Society of Microscopy [101]

The magnetic flux density B is generally estimated from the experimental reconstructed phase images using (2.31). Comparison of the experimental and simulated magnetic flux densities at the white circle in Fig. 3.77a estimated the effective value of thickness t at the position of the white circle to be 1.95×10^{-5} m. This value of t was used to estimate the magnetic flux densities from the distance between the contour lines l in the reconstructed phase images when various currents were applied. Figure 3.78 shows the experimental and calculated magnetic flux densities plotted as a function of the applied current. The measured values were in good agreement

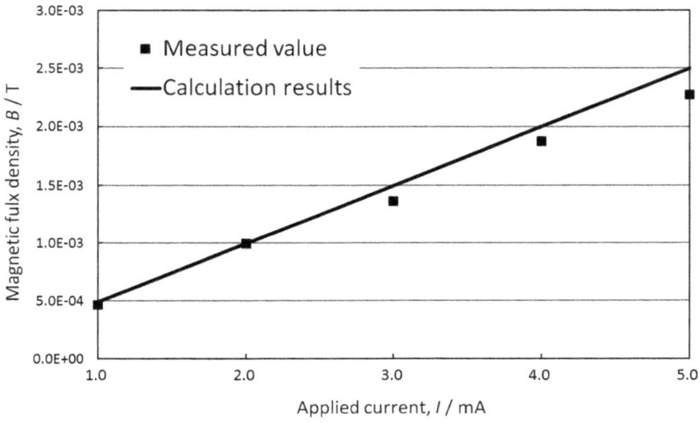

Fig. 3.78 Experimental and simulated magnetic flux densities at position of white circle in Fig. 3.77a plotted as a function of applied electric current. Permission obtained from The Japanese Society of Microscopy [101]

with the simulation results. The slight difference between the experimental result and the simulation is due to the influences of the inaccuracy of the model size of the magnetic pole.

The position of the magnetic pole cannot be located directly, because the magnetic pole is wholly embedded in the alumina part. Thus, for clarifying the magnetic field around the pole, another magnetic head was prepared for direct observation of the magnetic field around the magnetic pole. The magnetic head had a simple structure that consisted of alumina without magnetic shields, and the alumina parts around the magnetic pole were etched with alkaline solution for the TEM observation. Figure 3.79a shows an SEM image around the magnetic pole after etching, where the arrow indicates the location of the magnetic pole tip. The substrate from the etched area to the right and left edges of the magnetic recording head was removed by FIB processing for the TEM observation. Figure 3.79b shows a low magnification side view TEM image of the magnetic pole, where the arrow indicates the location of the magnetic pole tip. Figure 3.80a shows a hologram around the magnetic pole with no applied electric current, and Fig. 3.80b shows the reconstructed phase image of Fig. 3.80a. The spacing of the interference fringes was 2 nm and the field of view was approximately 350 nm. The phase information for the reconstructed phase image was subtracted from the reconstructed phase images obtained with an applied electric current. Figure 3.80c–e show reconstructed phase images around the magnetic pole with applied currents of 10, 20, and 30 mA, respectively. It should be noted that the phase shifts are wrapped by the cosine function. The magnetic flux gradually increases in the vicinity of the pole tip with an increase in the electric current. This experiment provide direct visualization of the magnetic field generated by a magnetic recording head. Therefore, it is expected that by utilizing electron holography, the magnetic field around the magnetic pole can be quantitatively evaluated with this

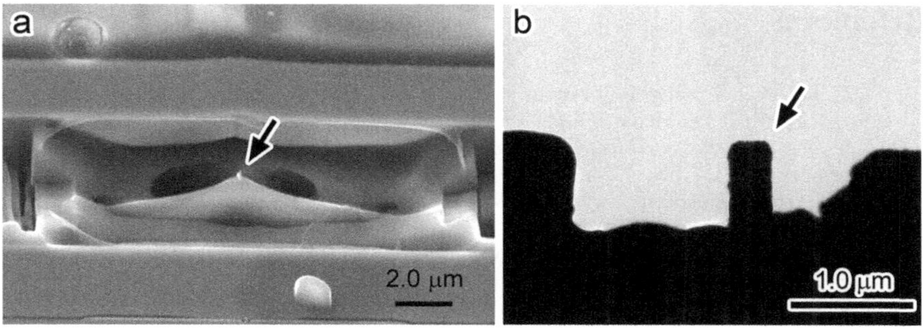

Fig. 3.79 **a** SEM and **b** transmission electron microscope images of magnetic pole. Region around magnetic pole was chemically etched for observation to allow location of pole tip. Permission obtained from The Japanese Society of Microscopy [101]

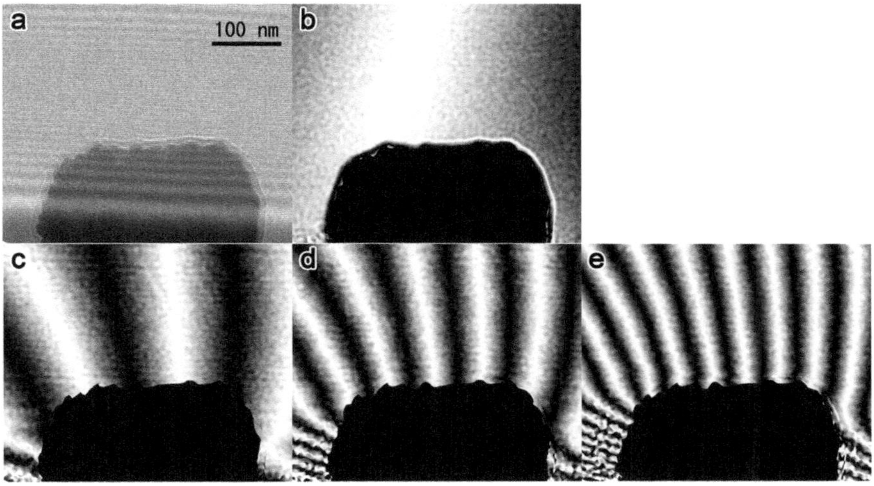

Fig. 3.80 **a** Hologram around magnetic pole without electric current. **b** Reconstructed phase image obtained from (**a**). **c–e** Reconstructed phase images with application of 10, 20, and 30 mA to magnetic recording head. Images of **c–e** were obtained by subtracting phase information of (**b**). Permission obtained from The Japanese Society of Microscopy [101]

technique and discussed in detail with respect to simulations that take into account the shape and characteristics of the magnetic recording head material.

References

1. D.C. Joy, A.D. Roming, J. Goldstein, *Principles of Analytical Electron Microscopy* (Springer Science & Business Media, 1986)
2. L. Reimer, *Transmission Electron Microscopy: Physics of Image Formation and Microanalysis* (Springer, 1984)
3. R.F. Egerton, *Electron Energy-Loss Spectroscopy in the Electron Microscope*, 2nd edn. (Plenum, New York, 1996)
4. D.B. Williams, C.B. Carter, *Transmission Electron Microscopy A Textbook for Materials Science* (Springer, 1996)
5. P.M. Kelly, A. Jostsons, R.G. Blake, J.G. Napier, Phys. Stat. Sol. A **31**, 771 (1975)
6. S.M. Allen, Philos. Mag. A **43**, 325 (1981)
7. M. Tanaka, M. Terauchi, *Convergent-Beam Electron Diffraction* (JEOL JTD., Tokyo, 1985)
8. J.C.H. Spence, J.M. Zuo, *Electron Microdiffraction* (Plenum Press, New York, 1992)
9. D. Nishino, A. Nakafuji, J.-M. Yang, D. Shindo, ISIJ Int. **38**, 1369 (1998)
10. M. Keller, Z. Phys. **164**, 274 (1961)
11. R. Buhl, Z. Phys. **155**, 395 (1959)
12. H. Hoffmann, C. Jönsson, Z. Phys. **182**, 360 (1965)
13. H.K. Gaukler, R. Schwarzer, Optik **33**, 215 (1971)
14. H. Suzuki, Z. Akase, K. Niitsu, T. Tanigaki, D. Shindo, Microscopy **66**, 167 (2017)
15. T. Tanigaki, K. Sato, Z. Akase, S. Aizawa, H. Soon Park, T. Matsuda, Y. Murakami, D. Shindo, H. Kawase, Appl Phys Lett **104**, 131601 (2014)
16. T. Sato, N. Tsukida, M. Higo, H. Magara, Z. Akase, D. Shindo, N. Ohno, Mater. Trans. **60**, 2114 (2019)
17. D. Shindo, Z. Akase, Mater. Sci. Eng. R. Rep. **142**, 100564 (2020)
18. Z. Akase, H. Magara, T. Sato, Y. Iwasaki, K. Shimada, W. HipKaeo, P. Srison, W. Sirithanaphol, U. Rompsaithong, H. Kondo, in *Proc 20th Int. Microsc. Congr. IMC20* (Busan, 2023), p. #0376
19. Y. Iwasaki, Z. Akase, K. Shimada, K. Harada, D. Shindo, Microscopy, **72**, 455 (2023)
20. F. Houdellier, G.M. Caruso, S. Weber, M. Kociak, A. Arbouet, Ultramicroscopy **186**, 128 (2018)
21. T. Niermann, T. Wagner, M. Lehmann, Microsc. Microanal. **24**, 1874 (2018)
22. F. Houdellier, G.M. Caruso, S. Weber, M.J. Hÿtch, C. Gatel, A. Arbouet, Ultramicroscopy **202**, 26 (2019)
23. K. Soma, S. Konings, R. Aso, N. Kamiuchi, G. Kobayashi, H. Yoshida, S. Takeda, Ultramicroscopy **181**, 27 (2017)
24. A. Feist, N. Bach, N. Rubiano da Silva, T. Danz, M. Möller, K.E. Priebe, T. Domröse, J.G. Gatzmann, S. Rost, J. Schauss, S. Strauch, R. Bormann, M. Sivis, S. Schäfer, C. Ropers, Ultramicroscopy **176**, 63 (2017)
25. A. Nakamura, T. Shimojima, Y. Chiashi, M. Kamitani, H. Sakai, S. Ishiwata, H. Li, K. Ishizaka, Nano Lett. **20**, 4932 (2020)
26. M. Möller, J.H. Gaida, S. Schäfer, C. Ropers, Commun. Phys. **3**, 1 (2020)
27. T. Shimojima, A. Nakamura, X. Yu, K. Karube, Y. Taguchi, Y. Tokura, K. Ishizaka, Sci. Adv. **7**, eabg1322 (2021)
28. J.W. Lau, K.B. Schliep, M.B. Katz, V.J. Gokhale, J.J. Gorman, C. Jing, A. Liu, Y. Zhao, E. Montgomery, H. Choe, W. Rush, A. Kanareykin, X. Fu, Y. Zhu, Rev. Sci. Instrum. **91**, 021301 (2020)
29. S.A. Reisbick, M.-G. Han, C. Liu, Y. Zhao, E. Montgomery, C. Jing, V.J. Gokhale, J.J. Gorman, J.W. Lau, Y. Zhu, Ultramicroscopy **235**, 113497 (2022)
30. Y. Gao, D. Shindo, Y. Bao, K. Krishnan, Appl. Phys. Lett. **90**, 233105 (2007)
31. J. Biskupek, U. Kaiser, H. Lichte, A. Lenk, T. Gemming, G. Pasold, W. Witthuhn, J. Magn. Magn. Mater. **293**, 924 (2005)
32. Y. Takeno, Y. Murakami, T. Sato, T. Tanigaki, H.S. Park, D. Shindo, R.M. Ferguson, K.M. Krishnan, Appl. Phys. Lett. **105**, 183102 (2014)

33. T. Kasama, R.J. Harrison, N.S. Church, M. Nagao, J.M. Feinberg, R.E. Dunin-Borkowski, Phase Transit. **86**, 67 (2013)
34. M.-T. Chang, L.-J. Chou, C.-H. Hsieh, Y.-L. Chueh, Z.L. Wang, Y. Murakami, D. Shindo, Adv. Mater. **19**, 2290 (2007)
35. *ELF Co* http://www.elf.co.jp/
36. M. Chiba, S. Kobayashi, K. Noguchi, T. Murakami, J.A. Szpunar, J. Manjanna, J. Magn. Magn. Mater. **512**, 167012 (2020)
37. T. Tanigaki, T. Akashi, Y. Takahashi, T. Kawasaki, H. Shinada, in *Adv. Imaging Electron Phys.*, edited by P. W. Hawkes (Elsevier, 2016), pp. 69–125
38. K. Harada, A. Tonomura, Y. Togawa, T. Akashi, T. Matsuda, Appl. Phys. Lett. **84**, 3229 (2004)
39. K. Harada, Microscopy **62**, S3 (2013)
40. A. Tonomura, T. Matsuda, J. Endo, T. Arii, K. Mihama, Phys. Rev. B Condens. Matter **34**, 3397 (1986)
41. D. Shindo, Y. Murakami, J. Phys. Appl. Phys. **41**, 183002 (2008)
42. M.J. Hÿtch, R.E. Dunin-Borkowski, M.R. Scheinfein, J. Moulin, C. Duhamel, F. Mazaleyrat, Y. Champion, Phys. Rev. Lett. **91**, 257207 (2003)
43. K. Niitsu, Y. Liu, A.C. Booth, X. Yu, N. Mathur, M.J. Stolt, D. Shindo, S. Jin, J. Zang, N. Nagaosa, Y. Tokura, Nat. Mater. **21**, 305 (2022)
44. Z.L. Wang, Adv. Mater. **12**, 1295 (2000)
45. Z.L. Wang, R.P. Gao, W.A. de Heer, P. Poncharal, Appl. Phys. Lett. **80**, 856 (2002)
46. J.-M. Bonard, K.A. Dean, B.F. Coll, C. Klinke, Phys. Rev. Lett. **89**, 197602 (2002)
47. N. de Jonge, M. Allioux, M. Doytcheva, M. Kaiser, K.B.K. Teo, R.G. Lacerda, W.I. Milne, Appl. Phys. Lett. **85**, 1607 (2004)
48. Z. Xu, X.D. Bai, E.G. Wang, Z.L. Wang, Appl. Phys. Lett. **87**, 163106 (2005)
49. H.J. Li, W.G. Lu, J.J. Li, X.D. Bai, C.Z. Gu, Phys. Rev. Lett. **95**, 086601 (2005)
50. A. Buldum, J.P. Lu, Phys. Rev. Lett. **91**, 236801 (2003)
51. J. Cumings, A. Zettl, M.R. McCartney, J.C.H. Spence, Phys. Rev. Lett. **88**, 056804 (2002)
52. J.J. Kim, D. Shindo, Y. Murakami, W. Xia, L.-J. Chou, Y.-L. Chueh, Nano Lett. **7**, 2243 (2007)
53. G. Matteucci, G.F. Missiroli, M. Muccini, G. Pozzi, Ultramicroscopy **45**, 77 (1992)
54. T. Oikawa, J.J. Kim, T. Tomita, H.S. Park, D. Shindo, J. Electron Microsc. **56**, 171 (2007)
55. J. J. Kim, W. X. Xia, D. Shindo, T. Oikawa, T. Tomita, Mater Trans **48**, 2631 (2007)
56. K.H. Kim, Z. Akase, D. Shindo, N. Ohno, Y. Fujii, N. Terada, S. Ohno, Microsc. Microanal. **19**(Suppl 5), 54 (2013)
57. D. Shindo, J. Jung Kim, K. Hyun Kim, W. Xia, N. Ohno, Y. Fujii, N. Terada, S. Ohno, J. Phys. Soc. Jpn **78**, 104802 (2009)
58. D. Shindo, J.J. Kim, W. Xia, K.H. Kim, N. Ohno, Y. Fujii, N. Terada, S. Ohno, J. Electron Microsc. **56**, 1 (2007)
59. D. Shindo, T. Tanigaki, H.S. Park, Adv. Mater. **29**, 1602216 (2017)
60. D. Shindo, S. Aizawa, Z. Akase, T. Tanigaki, Y. Murakami, H.S. Park, Microsc. Microanal. **20**, 1015 (2014)
61. W. Xia, K. Hirata, K. Yanagisawa, Y. Ishida, H. Kasai, K. Yanagiuchi, D. Shindo, A. Tonomura, J. Electron Microsc. (Tokyo) **59**, 331 (2010)
62. W. Gerlach, O. Stern, Z. Für Phys. **9**, 349 (1922)
63. S. Chikazumi, *Physics of Ferromagnetism, 2nd Edition* (Clarendon Press, 1997)
64. D. Shindo, T. Sato, US Patent 11,067,649, B1 (2021)
65. T. Sato, K. Shimada, Z. Akase, H. Magara, T. Tomita, D. Shindo, Microscopy **73**, 367 (2024)
66. M. Inoue, T. Tomita, M. Naruse, Z. Akase, Y. Murakami, D. Shindo, J. Electron Microsc. **54**, 509 (2005)
67. J.D. Jackson, *Classical Electrodynamics* (Wiley, New York, London, 1962)
68. D. Shindo, T. Tomita, *Material Characterization Using Electron Holography* (Wiley-VCH, Weinheim, 2023)
69. D. Shindo, T. Musashi, Y. Ikematsu, Y. Murakami, N. Nakamura, H. Chiba, J. Electron Microsc. **54**, 11 (2005)

70. S.A. Chambers, L. Qiao, T.C. Droubay, T.C. Kaspar, B.W. Arey, P.V. Sushko, Phys. Rev. Lett. **107**, 206802 (2011)
71. M. Imada, A. Fujimori, Y. Tokura, Rev Mod Phys **70**, 1039 (1998)
72. T. Arima, Y. Tokura, J.B. Torrance, Phys. Rev. B **48**, 17006 (1993)
73. H.J. Zhao, Y. Yang, W. Ren et al., J. Phys. Condens. Matter **26**, 472201 (2014)
74. M. Nakamura, F. Kagawa, T. Tanigaki, H.S. Park, T. Matsuda, D. Shindo, Y. Tokura, M. Kawasaki, Phys. Rev. Lett. **116**, 156801 (2016)
75. E.G. Visser, J. Magn. Magn. Mater. **42**, 286 (1984)
76. K. Kondo, T. Chiba, S. Yamada, E. Otsuki, J. Appl. Phys. **87**, 6229 (2000)
77. T. Kasahara, D. Shindo, H. Yoshikawa, T. Sato, K. Kondo, J. Electron Microsc. (Tokyo) **56**, 7 (2007)
78. S.J. Hefferman, J.N. Chapman, S. McVitie, J. Magn. Magn. Mater. **95**, 76 (1991)
79. T. Uhlig, M. Heumann, J. Zweck, Ultramicroscopy **94**, 193 (2003)
80. G. Yi, W.A.P. Nicholson, C.K. Lim, J.N. Chapman, S. McVitie, C.D.W. Wilkinson, Ultramicroscopy **99**, 65 (2004)
81. J. Cumings, E. Olsson, A.K. Petford-Long, Y. Zhu, MRS Bull. **33**, 101 (2008)
82. Z. Akase, D. Shindo, J. Electron Microsc. (Tokyo) **59**, 207 (2010)
83. M. Arita, R. Tokuda, K. Hamada, Y. Takahashi, Mater. Trans. **55**, 403 (2014)
84. J. Gründmayer, J. Zweck, in *Proc 16th Int Microsc. Congr.* (Sapporo, 2006), p. 1541
85. Z. Akase, D. Shindo, M. Inoue, A. Taniyama, Mater. Trans. **48**, 2626 (2007)
86. Z. Akase, S. Aizawa, D. Shindo, P. Sharma, A. Makino, J. Magn. Magn. Mater. **375**, 10 (2015)
87. D. Shindo, Y.-G. Park, Y. Gao, H.S. Park, J. Appl. Phys. **95**, 6521 (2004)
88. H.S. Park, Y.-G. Park, Y. Gao, D. Shindo, M. Inoue, J. Appl. Phys. **97**, 033908 (2005)
89. Z. Akase, K. Shinmeimae, Y. Miyabayashi, T. Sato, D. Shindo, K. Niitsu, in *76th Annu. Meet. Jpn. Soc. Microsc. Osaka May 25–27* (Osaka, 2020)
90. Z. Akase, M. Higo, K. Shimada, T. Sato, H. Magara, D. Shindo, N. Ohno, Mater. Trans. **62**, 1589 (2021)
91. A.M. Campbell, J.E. Evetts, Adv. Phys. **21**, 199 (1972)
92. J. Plain, T. Puig, F. Sandiumenge, X. Obradors, J. Rabier, Phys. Rev. B Condens. Matter Mater. Phys. **65**, 104526 (2002)
93. Z. Akase, H. Kasai, S. Mamishin, D. Shindo, M. Morita, A. Tonomura, J. Appl. Phys. **111**, 033912 (2012)
94. M. Morita, M. Sawamura, S. Takebayashi, K. Kimura, H. Teshima, M. Tanaka, K. Miyamoto, M. Hashimoto, Phys. C Supercond **235–240**, 209 (1994)
95. M. Morita, M. Kumada, A. Sato, H. Teshima, H. Hirano, Phys C **463–465**, 387 (2007)
96. K. Harada, J. Endo, N. Osakabe, A. Tonomura, E.-J. Surf, Sci. Nanotechnol. **6**, 29 (2008)
97. K.H. Kim, Z. Akase, T. Suzuki, D. Shindo, Mater. Trans. **51**, 1080 (2010)
98. H. Lichte, Ultramicroscopy **108**, 256 (2008)
99. D. Shindo, K. Takahashi, Y. Murakami, K. Yamazaki, S. Deguchi, H. Suga, Y. Kondo, J. Electron Microsc. (Tokyo) **58**, 245 (2009)
100. H. Okada, D. Shindo, J.J. Kim, Y. Murakami, H. Kawase, J. Appl. Phys. **102**, 054908 (2007)
101. T. Goto, J.S. Jeong, W. Xia, Z. Akase, D. Shindo, K. Hirata, Microscopy **62**, 383 (2013)
102. JRI Solutions, Ltd., https://www.jmag-international.com/
103. Y. Aoyama, Y.-G. Park, D. Shindo, Microscopy **54**, 279 (2005)
104. G. Matteucci, G. Missiroli, G. Pozzi, Scanning Microsc. **11**, 367 (1997)

Chapter 4
Dynamical Diffraction and Strong Deflection

Abstract Effects of dynamical diffraction and strong deflection on electron holography analysis are explained. Phase contrast variations due to the dynamical diffraction effect are explained on the basis of computer simulation results. The characteristic features of the strong deflection effect on reconstructed phase images are explained in terms of the wave-particle dualism of the electron, which is considered in Chap. 5.

In this chapter, complicated problems in electron holography analysis, i.e., the dynamical diffraction effect and strong deflection in the visualization of electromagnetic fields in and around various electromagnetic materials are described and several examples of their interpretation and treatments are presented.

4.1 Dynamical Diffraction

4.1.1 Outlines of Kinematical and Dynamical Diffraction Theories

First, the bases of kinematical and dynamical theories of electron diffraction and the effect of dynamical diffraction on the phase modulation of electron beams are explained. In the case of diffraction phenomena, the kinematical theory considers only the diffracted waves scattered once in a specimen; by contrast, the dynamical theory considers multiple scattering. In electron diffraction, dynamical theory is necessary to interpret the scattering intensity accurately because of the large interaction between electrons and the electric potential of the specimen. The effect of dynamical diffraction is also apparent in the phase modulation in electron beams in electron holography.

In the following sections, we introduce the kinematical theory of electron diffraction in transmission electron microscopy and then describe three representative

D. Shindo and Z. Akase, *Electron Holography for Electromagnetic Materials*, Springer Series in Materials Science 347, https://doi.org/10.1007/978-981-97-3537-2_4

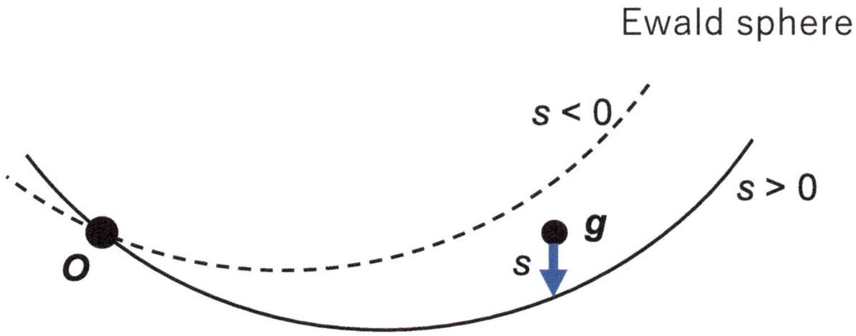

Fig. 4.1 Illustration showing sign of *s*. *s* is positive when reflection is inside Ewald sphere

methods of implementing the dynamical theory of electron diffraction: (i) the method of solving the Howie–Whelan differential equations based on wave optics, (ii) the Bethe method starting from the Schrödinger equation and solving matrix eigenvalue problems, and (iii) the multi-slice method, which can calculate the image and diffraction intensities sequentially in the specimen thickness direction.

4.1.2 Kinematical Theory

In the kinematical theory of electron diffraction, the probability of the amplitude of a diffracted wave is calculated by summing the probability amplitudes of electrons scattered once in the specimen over all points of the specimen. In practice, the sum of the contributions in the plane perpendicular to the depth of the specimen is calculated first and is then integrated over the depth direction. In the case of a specimen for transmission electron microscopy, the electron contribution is mainly from inside the column, with a spread of ~1 nm; the contributions from outside the column almost cancel each other because of the strong phase oscillation. Therefore, we consider only the contribution from inside the column. The sum of contributions to the *g*-reflection from a plane at depth *z* inside the column is given by

$$\frac{\mathrm{d}\psi_g}{\mathrm{d}z} = \frac{i\pi}{\xi_g}\exp(-2\pi isz), \tag{4.1}$$

where ξ_g is the extinction distance ($\xi_g = \pi V_c \cos\theta_B/(\lambda F_g)$, where V_c is the volume of the unit cell, θ_B is the Bragg angle, and F_g is the crystal structure factor) and s is the excitation error corresponding to the distance between the reciprocal lattice point *g* and the Ewald sphere in reciprocal space (Fig. 4.1). In the kinematical theory of diffraction, the wave function ψ_g of the *g* diffraction wave is the sum of the contributions from each surface of the specimen from the top ($z = 0$) to the bottom ($z = t$):

$$\psi_g = \frac{i\pi}{\xi_g} \int_0^t \exp(-2\pi isz)\mathrm{d}z = \frac{i\pi}{\xi_g} \frac{\sin(\pi st)}{\pi s} \exp(-\pi ist). \tag{4.2}$$

Thus, the intensity of the g-reflection is given by

$$I_g = {\psi_g}^* \psi_g = \left(\frac{\pi}{\xi_g}\right)^2 \frac{\sin^2(\pi st)}{(\pi s)^2}, \tag{4.3}$$

where the superscript * indicates a complex conjugate. The diffraction intensity I_g is proportional to the square of the crystal structure factor F_g (i.e., $\xi_g \propto 1/F_g$) and oscillates with a period of $1/s$ with respect to the specimen thickness. Numerous other features of electron diffraction intensity and image contrast can also be explained by the kinetical theory of diffraction. However, there are limitations to its application. For example, when $s = 0$, i.e., when the Bragg condition is satisfied ($I_g \rightarrow (\pi t/\xi_g)^2$), the diffraction intensity oscillates with respect to specimen thickness in experiments, whereas it increases monotonically in calculations.

4.1.3 Dynamical Theories with Bloch Wave Treatment

Dynamical Theory Based on Howie–Whelan Differential Equations

In the case of multiple scattering, it is necessary to consider the product of the probability amplitudes, which means that an electron scatters at one point in the specimen and then scatters again at a different point. Here, we consider multiple scattering in the two-wave excitation condition, where the direct wave and only one diffracted wave of the g-reflection are excited.

The contribution $\mathrm{d}\psi_0/\mathrm{d}z$ to the g-reflection from a plane at depth z in the column is given by

$$\frac{\mathrm{d}\psi_g}{\mathrm{d}z} = \frac{i\pi}{\xi_g} \psi_0 \exp(-2\pi isz) + \frac{i\pi}{\xi_0} \psi_g. \tag{4.4}$$

The first term on the right-hand side is the component that scatters from the direct wave and contributes to the g-diffracted wave, as in (4.1); however, the difference from (4.1) is that this term is the product of the probability amplitude ψ_0 of the transmitted wave at depth z. The second term on the right-hand side is the product of the probability amplitude ψ_g of the g-diffracted wave at depth z and the probability amplitude $(i\pi/\xi_0)$ of the diffracted wave transmitting forward. Similarly, the contribution $\mathrm{d}\psi_0/\mathrm{d}z$ to the direct wave can be written as

$$\frac{\mathrm{d}\psi_0}{\mathrm{d}z} = \frac{i\pi}{\xi_0} \psi_0 + \frac{i\pi}{\xi_g} \psi_g \exp(2\pi isz). \tag{4.5}$$

The simultaneous differential equations in (4.4 and 4.5) are called the Howie–Whelan equations. Simultaneous differential equations of this form were originally introduced by Darwin in 1914 for X-ray diffraction analysis [1] and were later applied to electron diffraction by Howie and Whelan in 1961 [2]. These simultaneous equations can be transformed into the following second-order linear ordinary differential equation by substitution of variables rearranging $\psi_0\exp(-\pi iz/\xi_0)$ as ψ_0 and $\psi_g\exp(2\pi isz - \pi iz/\xi_0)$ as ψ_g:

$$\frac{d^2\psi_0}{dz^2} - 2\pi is\frac{d\psi_0}{dz} + \frac{\pi^2}{\xi_g{}^2}\psi_0 = 0. \tag{4.6}$$

Because this equation is a second-order linear ordinary differential equation, ψ_0 has a special solution of the form $\psi_0 = \exp(2\pi i\gamma z)$ and the above equation can be transformed into a second-order equation (characteristic equation) of γ. The solutions of the characteristic equation are

$$\gamma^{(1)} = \frac{\left(s+\sqrt{s^2+\frac{1}{\xi_g{}^2}}\right)}{2},$$
$$\gamma^{(2)} = \frac{\left(s-\sqrt{s^2+\frac{1}{\xi_g{}^2}}\right)}{2}. \tag{4.7}$$

Therefore, the general solution of (4.6) is

$$\psi_0 = C_0^{(1)}\exp\left(2\pi i\gamma^{(1)}z\right) + C_0^{(2)}\exp\left(2\pi i\gamma^{(2)}z\right), \tag{4.8}$$

which is a linear combination of the special solutions. The coefficient $C_0{}^{(j)}$ is an arbitrary constant at this point. Similarly, for the diffracted electron,

$$\psi_g = C_g^{(1)}\exp\left(2\pi i\gamma^{(1)}z\right) + C_g^{(2)}\exp\left(2\pi i\gamma^{(2)}z\right), \tag{4.9}$$

where there is the relation $C_g{}^{(j)}/C_0{}^{(j)} = 2\xi_g\gamma^{(j)}$ between $C_g{}^{(j)}$ and $C_0{}^{(j)}$. We can now normalize C_g and C_0 to impose a restriction on the absolute magnitudes of direct and diffracted waves:

$$C_0{}^{(1)2} + C_g{}^{(1)2} = C_0{}^{(2)2} + C_g{}^{(2)2} = 1. \tag{4.10}$$

Using C and γ obtained above, we can describe the following two independent solutions (Bloch waves) as wave functions that satisfy the Howie–Whelan equations:

$$b^{(1)}\left(\vec{k}^{(1)}, \vec{r}\right) = C_0{}^{(1)}\exp\left(2\pi i\vec{k}^{(1)}\cdot\vec{r}\right) + C_g{}^{(1)}\exp\left(2\pi i(\vec{k}^{(1)} + \vec{g})\cdot\vec{r}\right)$$
$$b^{(2)}\left(\vec{k}^{(2)}, \vec{r}\right) = C_0{}^{(2)}\exp\left(2\pi i\vec{k}^{(2)}\cdot\vec{r}\right) + C_g{}^{(2)}\exp\left(2\pi i\left(\vec{k}^{(2)} + \vec{g}\right)\cdot\vec{r}\right). \tag{4.11}$$

Let $\vec{K}$ be the wavenumber vector of the incident electrons in the crystal including the effect of refraction, and let the z component of $(\vec{k} - \vec{K})$ be γ. The wave function Ψ of the electrons in the crystal is a linear combination of Bloch wave (1) and Bloch wave (2):

$$\Psi = \alpha^{(1)} b^{(1)} + \alpha^{(2)} b^{(2)}. \tag{4.12}$$

The Bloch wave excitation coefficients $\alpha^{(1)}$ and $\alpha^{(2)}$ are determined from the boundary conditions (i.e., the transmitted wave intensity is 1 and the diffracted wave intensity is 0 at the top surface of the specimen). When two parameters w and β are defined as

$$w = s\xi_g, \ \cot\beta = w, \tag{4.13}$$

C and α can be written as

$$C_0{}^{(1)} = \sin\tfrac{\beta}{2}, \ C_g{}^{(1)} = \cos\tfrac{\beta}{2}, \ C_0{}^{(2)} = \cos\tfrac{\beta}{2}, \ C_g{}^{(2)} = -\sin\tfrac{\beta}{2}$$

$$\alpha^{(1)} = \sin\tfrac{\beta}{2}, \ \alpha^{(2)} = \cos\tfrac{\beta}{2}. \tag{4.14}$$

When the g-reflection satisfies the Bragg reflection condition, $s = 0$, $w = 0$, and $\beta = \pi/2$. When the reflection angle θ is larger than the Bragg angle θ_B, $s > 0$, $w > 0$, and $\beta < \pi/2$.

Finally, the g diffraction intensity I_g takes a form similar to (4.3) obtained by the kinematic diffraction theory, as shown in the following equation:

$$I_g = \psi_g{}^* \psi_g = \frac{1}{w^2+1}\sin^2\frac{\pi t\sqrt{w^2+1}}{\xi_g} = \left(\frac{\pi}{\xi_g}\right)^2 \frac{\sin^2(\pi s_{\text{eff}} t)}{(\pi s_{\text{eff}})^2}, \tag{4.15}$$

where s_{eff} is the effective excitation error,

$$s_{\text{eff}} = \sqrt{s^2 + \frac{1}{\xi_g{}^2}} = \frac{\sqrt{w^2+1}}{\xi_g}. \tag{4.16}$$

The inverse of the effective excitation error s_{eff} is called the effective extinction distance $\xi_{\text{g,eff}}$. In (4.15), when $s = 0$, which satisfies the Bragg condition, the g diffraction intensity oscillates with a period of the extinction distance ξ_{g} with respect to the specimen thickness.

A diagram depicting the situation in (4.11) in reciprocal space in the manner the Ewald construction is shown in Fig. 4.2. The wave shown by the combination of red wavenumber vectors in the figure ($\vec{k}^{(1)}$ and $\vec{k}^{(1)} + \vec{g}$) is Bloch wave (1), and the wave shown by the combination of blue wavenumber vectors ($\vec{k}^{(2)}$ and $\vec{k}^{(2)} + \vec{g}$) is Bloch wave (2). The starting point of the wavenumber vector $\vec{K}$ of the incident electron is located on a sphere of radius K from the origin O, whereas the starting

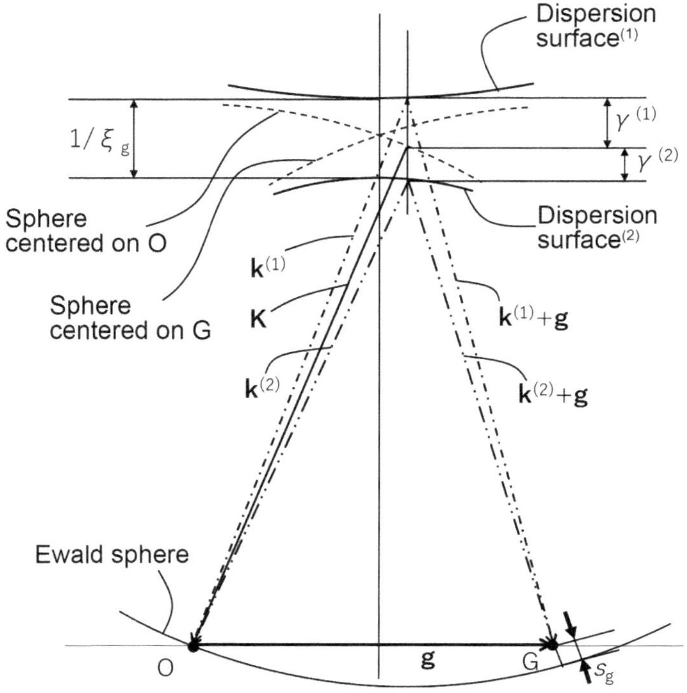

Fig. 4.2 Schematic of Bloch waves and dispersion surfaces

point of the vector of the Bloch wave $\overrightarrow{k}$ is displaced by γ from that sphere. The plane connecting the starting points of the wavenumber vectors of Bloch waves for various incident directions is called the dispersion surface. Where the Bragg condition is satisfied, the distance between dispersion surfaces is the reciprocal of the extinction distance ($1/\xi_g$). This figure shows the two-wave excitation state of a direct wave and a diffracted wave; however, when n diffracted waves are considered in the multi-wave dynamical diffraction theory described below, n dispersive surfaces also arise and n Bloch waves are considered. In electron diffraction, the intensities of the direct and diffracted waves oscillate with respect to the thickness of the specimen, which can be interpreted as the result of the addition of Bloch waves with slightly different wavenumbers (i.e., different wavelengths), resulting in beating.

Dynamical Theory Based on the Bethe Method

In this section, the Bethe method, one of the dynamical diffraction theories, is outlined. The Bethe method is a method for determining the motion of electrons in a crystal potential φ from the Schrödinger equation, first conceived by Bethe

[3] and later developed by MacGillavry [4], Heidenreich [5], Kato [6], Sturkey [7], Fujimoto [8], and Niehrs [9, 10].

The electron wave function Ψ obeys Schrödinger's wave equation:

$$\nabla^2\Psi(\vec{r}) + \tfrac{8\pi^2 me}{h^2}\left(V + \varphi(\vec{r})\right)\Psi(\vec{r}) = 0, \tag{4.17}$$

where m is the electron mass, e is elementary electrical charge, h is Planck's constant, and V is the acceleration voltage. $\varphi(\vec{r})$ is the crystal potential, which can be described by a Fourier series as follows:

$$\begin{aligned}\varphi(\vec{r}) &= \sum_h V_h \exp(2\pi i \vec{h}\cdot\vec{r}) \\ &= \frac{h^2}{2me}\left(\sum_h U_h \exp\left(2\pi i \vec{h}\cdot\vec{r}\right)\right) \\ &= \tfrac{h^2 U_0}{2me} + \tfrac{h^2}{2me}\left(\sum_{h\neq 0} U_h \exp\left(2\pi i \vec{h}\cdot\vec{r}\right)\right),\end{aligned} \tag{4.18}$$

where $\vec{h}$ is the reciprocal lattice vector. The crystal structure factor U_g, which is the Fourier coefficient of $\varphi(\vec{r})$, can be calculated as

$$U_g = \tfrac{2m|e|}{h^2 V_c}\sum_j f_j^e \exp(-2\pi i\vec{g}\cdot\vec{r}_j)\exp\left(\tfrac{-B_j|\vec{g}|^2}{4}\right), \tag{4.19}$$

where f_j^e is the atomic scattering factor for the j-th atom and the summation is carried out for each atom in the unit cell, $\exp\left(-B_j|\vec{g}|^2/4\right)$ is called the temperature factor and is used to account for wave damping due to thermal vibration of the atoms, and B_j is called the isotropic atomic displacement parameter.

The Schrödinger equation of (4.17) has the Bloch wave given by the following equation as a special solution:

$$\Psi(\vec{r}) = \sum_g C_g\left(\vec{k}\right)\exp\left\{2\pi i\left(\vec{k}+\vec{g}\right)\cdot\vec{r}\right\}. \tag{4.20}$$

Substituting (4.18) for the crystal potential and (4.20) for the Bloch wave into (4.17), and further assuming the following equation

$$K^2 = \tfrac{2meE}{h^2} + U_0, \tag{4.21}$$

Equation 4.17 can then be organized in exponential terms:

$$\sum_{g}\left[\left\{K^2-\left(\vec{k}+\vec{g}\right)^2\right\}C_g\left(\vec{k}\right)+\sum_{h\neq 0}U_hC_{g-h}\right]\exp\left\{2\pi i\left(\vec{k}+\vec{g}\right)\cdot\vec{r}\right\}=0. \tag{4.22}$$

For this expression to hold for any r, the coefficients of the exponents must all be zero. Thus, we obtain a series of simultaneous equations shown in the following equation for a series of $\vec{g}$:

$$\left\{K^2-\left(\vec{k}+\vec{g}\right)^2\right\}C_g\left(\vec{k}\right)+\sum_{h\neq 0}U_hC_{g-h}=0. \tag{4.23}$$

These simultaneous equations are the fundamental equations of the dynamical theory of multi-beam electron diffraction. Originally, there are an infinite number of g and h; however, in actual calculations, they are terminated at a finite number depending on the required calculation accuracy. In matrix notation, (4.23) can be expressed as

$$\begin{pmatrix} \left|\vec{K}\right|^2-\left|\vec{k}\right|^2 & U_{-g} & U_{-h} & \cdots \\ U_g & \left|\vec{K}\right|^2-\left|\vec{k}+\vec{g}\right|^2 & U_{g-h} & \cdots \\ U_h & U_{h-g} & \left|\vec{K}\right|^2-\left|\vec{k}+\vec{h}\right|^2 & \cdots \\ \vdots & \vdots & \vdots & \ddots \end{pmatrix}\begin{pmatrix} C_0 \\ C_g \\ C_h \\ \vdots \end{pmatrix}=0. \tag{4.24}$$

Here, from the dispersion surface diagram in Fig. 4.2, the following approximation is possible:

$$\left|\vec{k}+\vec{g}\right|\approx K+\gamma-s_g. \tag{4.25}$$

The diagonal elements of the matrix in (4.24) then become

$$\left|\vec{K}\right|^2-\left|\vec{k}+\vec{g}\right|^2\approx 2K\left(s_g-\gamma\right). \tag{4.26}$$

Therefore, (4.24) can be transformed as

$$\begin{pmatrix} 0 & \frac{U_{-g}}{2K} & \frac{U_{-h}}{2K} & \cdots \\ \frac{U_g}{2K} & s_g & \frac{U_{g-h}}{2K} & \cdots \\ \frac{U_h}{2K} & \frac{U_{h-g}}{2K} & s_h & \cdots \\ \vdots & \vdots & \vdots & \ddots \end{pmatrix}\begin{pmatrix} C_0 \\ C_g \\ C_h \\ \vdots \end{pmatrix}=\gamma\begin{pmatrix} C_0 \\ C_g \\ C_h \\ \vdots \end{pmatrix}. \tag{4.27}$$

The matrix on the left-hand side is a square matrix containing excitation errors (information about crystal orientation) in the diagonal elements and crystal structure factors (information about crystal structure) in the off-diagonal elements. Equation 4.27 is attributed to the eigenvalue problem, where the eigenvalues of this square matrix are γ and the eigenvectors are C_g. Parameters γ and C_g can be solved numerically using scientific calculation subroutines for eigenvalue problems. To consider the intensity decay due to inelastic scattering, we can effectively incorporate the crystal structure factor in the above matrix by replacing $U_g \to U_g + iU_g{}'$ and adding an imaginary component [11–14] (the amplitude of $U_g{}'$ is set to less than 1/10 of U_g [11]). The eigenvalues then become complex, and the imaginary components of the eigenvalues play the role of intensity decay with respect to sample thickness.

Next, the excitation coefficients of the Bloch waves, $\alpha^{(j)}$, are obtained by the boundary conditions. The equation for a g diffraction wave at thickness t is

$$\psi_g = \sum_{j=1}^{n} \alpha^{(j)} C_g^{(j)} \exp\left(2\pi i \gamma^{(j)} t\right), \tag{4.28}$$

where the phase term exp(2πiKt), which has no effect on the intensity calculation, is omitted. Equation 4.28 for each g can be expressed collectively in matrix form as

$$\begin{pmatrix} \psi_0(t) \\ \psi_g(t) \\ \psi_h(t) \\ \vdots \end{pmatrix} =$$

$$\begin{pmatrix} C_0^{(1)} & C_0^{(2)} & C_0^{(3)} & \cdots \\ C_g^{(1)} & C_g^{(2)} & C_g^{(3)} & \cdots \\ C_h^{(1)} & C_h^{(2)} & C_h^{(3)} & \cdots \\ \vdots & \vdots & \vdots & \ddots \end{pmatrix} \begin{pmatrix} \exp\left(2\pi i\gamma^{(1)}t\right) & 0 & 0 & \cdots \\ 0 & \exp\left(2\pi i\gamma^{(2)}t\right) & 0 & \cdots \\ 0 & 0 & \exp\left(2\pi i\gamma^{(3)}t\right) & \cdots \\ \vdots & \vdots & \vdots & \ddots \end{pmatrix} \begin{pmatrix} \alpha^{(1)} \\ \alpha^{(2)} \\ \alpha^{(3)} \\ \vdots \end{pmatrix}. \tag{4.29}$$

Let **C** be the matrix consisting of the eigenvectors on the left- of the right-hand sides, and consider the boundary conditions that the transmitted wave intensity is 1 and the diffracted wave intensity is 0 at the top surface of the sample (i.e., $t = 0$, ψ_0 $(t = 0) = 1$, ψ_g $(t = 0) = 0$, ψ_h $(t = 0) = 0$, …). Substituting these conditions into (4.29), we obtain

$$\begin{pmatrix} 1 \\ 0 \\ 0 \\ \vdots \end{pmatrix} = C \begin{pmatrix} \alpha^{(1)} \\ \alpha^{(2)} \\ \alpha^{(3)} \\ \vdots \end{pmatrix}. \tag{4.30}$$

Therefore, by multiplying both sides by the inverse of **C**, the excitation coefficient of the Bloch wave $\alpha^{(j)}$ can be calculated as

$$\begin{pmatrix} \alpha^{(1)} \\ \alpha^{(2)} \\ \alpha^{(3)} \\ \vdots \end{pmatrix} = \mathbf{C}^{-1} \begin{pmatrix} 1 \\ 0 \\ 0 \\ \vdots \end{pmatrix}. \tag{4.31}$$

4.1.4 Dynamical Theory Based on Multi-slice Method

The multi-slice method was proposed by Cowley and Moodie [15–17]. As described below, this method starts by creating a projection potential in real space; unlike the Bethe method, which starts with a perfect crystal, it can easily handle the propagation of electron waves in complex models such as grain boundaries and is often used to interpret high-resolution electron microscopy images.

The crystal is divided into n slices, and the thickness of one slice is Δz. The electron beam is assumed to be nearly perpendicular to this slice. Initially, we consider that scattering by the electric potential φ occurs at the first slice located on the top surface of the crystal, which causes the electron wave to change phase as represented in the following equation (transmission function):

$$q(x, y) = \exp\{i\sigma\varphi(x, y)\Delta z - \mu(x, y)\Delta z\}, \tag{4.32}$$

where σ is the interaction constant (2.16) and $\mu\Delta z$ is the effect of absorption. Next, we consider that the electron beam scattered at the upper surface of the first slice at small angles propagates in vacuum to the lower surface of the first slice (Fig. 4.3). This scattering process is described by the propagation function:

$$p(x, y) = \frac{1}{i\Delta z\lambda}\exp\left\{\frac{ik\left(x^2+y^2\right)}{2\Delta z}\right\}. \tag{4.33}$$

The scattering amplitude ψ_1 at the lower surface of the first slice can be expressed by

$$\psi_1(x, y) = q(x, y) \otimes p(x, y), \tag{4.34}$$

where $\otimes$ is the convolution operator. To calculate the scattering amplitude ψ_2 of the lower surface of the second slice, consider that the incident wave ψ_1 scatters on the upper surface of the second slice and then propagates to the lower surface of the second slice; that is,

$$\psi_2(x, y) = (q(x, y)\psi_1(x, y)) \otimes p(x, y)$$

$$= q(x, y)\{q(x, y) \otimes p(x, y)\} \otimes p(x, y). \tag{4.35}$$

Using the following similar operation, we can express the scattering amplitude ψ_n at the lower surface of the specimen consisting of n slices as

$$\psi_n(x, y) = q(x, y) \underset{n-1}{[} \cdots \underset{2}{[} q(x, y) \underset{1}{[} q(x, y) \otimes p(x, y) \underset{1}{]} \otimes p(x, y) \underset{2}{]} \cdots \underset{n-1}{]} \otimes p(x, y), \tag{4.36}$$

which can be calculated sequentially. This formula is based on real space; however, the operation can also be performed in reciprocal space using the Fourier transformation. Fourier transformation of the above equation yields

$$\psi_n(u, v) = \underset{n-1}{[} Q(u, v) \otimes \cdots \underset{2}{[} Q(u, v) \otimes \underset{1}{[} Q(u, v) \otimes Q(u, v)P(u, v) \underset{1}{]} P(u, v) \underset{2}{]} \cdots \underset{n-1}{]} P(u, v), \tag{4.37}$$

where $Q(u, v)$ and $P(u, v)$ are the functions obtained by Fourier transformation of $q(x, y)$ and $p(x, y)$, respectively. When the specimen is a crystal, $Q(u, v)$ reaches a maximum when (u, v) coincides with the reciprocal lattice vector and is discrete in the reciprocal space; thus, the convolution and product in (4.37) need only be carried out at the discrete reciprocal points. In the reciprocal space, the propagation function $P(u, v)$ is

$$P(u, v) = \exp\{-i\pi\lambda\Delta z(u^2 + v^2)\}. \tag{4.38}$$

If the incident beam is inclined from the crystal band axis by an angle (α_x, α_y), the effect can be taken into account by multiplying (4.38) by the following equation:

$$\exp\{2\pi i\Delta z(u\tan\alpha_x + v\tan\alpha_y)\}. \tag{4.39}$$

The flow chart of the multi-slice calculation for high-resolution electron microscope images is presented in Fig. 4.4.

Some examples of the use of the dynamical diffraction effect are briefly presented. First, a simple application derived from (4.15) is to estimate the specimen thickness and extinction distance from the rocking curve in a reflection disk of convergent-beam electron diffraction in a systematic row [18, 19]. Other applications of the Bethe method including studying the precise determination of crystal structure factors from the intensity of convergent-beam electron diffraction patterns. One of the applications is the critical voltage method [20], which uses an anomaly in the diffraction intensity at the point where γ has a multiple root when the acceleration voltage is varied [21–23]. The literature also includes studies that directly compare diffraction intensities with simulation results. Given that the Friedel rule is broken in dynamical diffraction

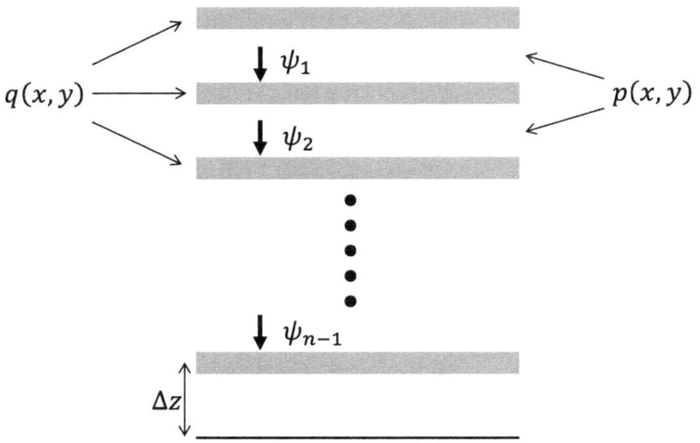

Fig. 4.3 Transmission function $q(x, y)$ and propagation function $p(x, y)$ in multi-slice method

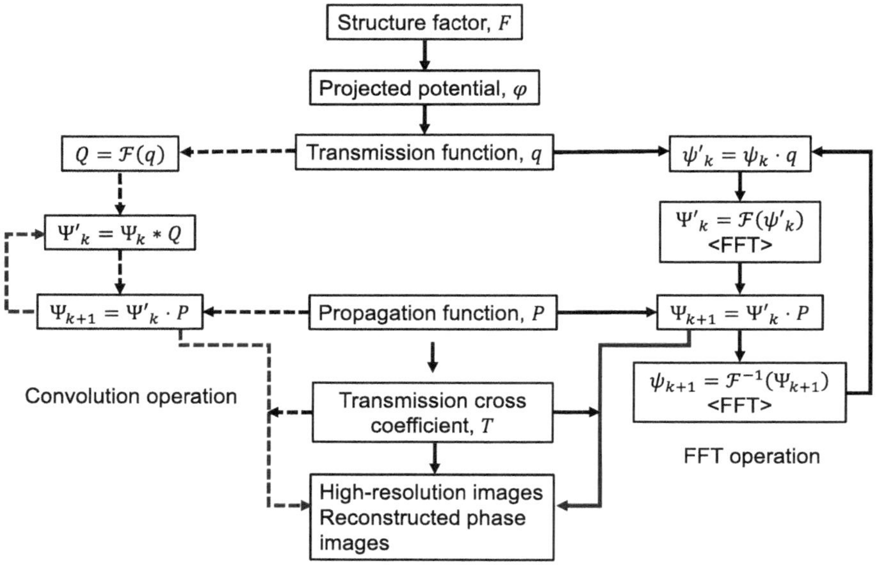

Fig. 4.4 Flow chart of dynamical calculation program based on multi-slice method

[24–26], the space group of a crystal from the symmetry of a convergent-beam electron diffraction pattern can be determined. The atom location channeling enhanced microanalysis (ALCHEMI) method [27–30], which combines electron channeling and EDS to determine the occupied sites of solid-solution atoms, is another technique based on the dynamical diffraction effect.

For high-resolution electron microscopy and electron holography at the atomic scale, the incident electron beam tends to be parallel to the zone axis. Under such an experimental condition, except in the case of very thin specimens, the dynamical diffraction effect is enhanced because many diffraction spots are excited. To interpret the high-resolution images and the reconstructed phase images under this experimental condition, the dynamical diffraction effect should be taken into account using the aforementioned multi-slice method. In the following chapters, we discuss the dynamical diffraction effect on the reconstructed phase images obtained under the condition where a limited number of diffraction spots are excited. For the interpretation of the reconstructed phase images, the dynamical diffraction calculation based on the Bethe method is used.

4.1.5 Interpretation of Dynamic Diffraction Effects in Electron Holography

According to the dynamical theory of electron diffraction, the electron wave in a crystal field is represented by a linear combination of multiple Bloch waves with slightly different wavelengths and the excitation ratio for each Bloch wave is determined by the angle of incidence of the electron beam on the crystal. Because the electron wave is the sum of Bloch waves with different wavelengths, the intensity of the electron beam produces a beat in the direction of the specimen thickness, resulting in thickness fringes in a bright-field image. In addition, depending on the diffraction conditions, especially those in the vicinity of the Bragg conditions, the phase does not always vary monotonically with specimen thickness. A frequently observed phenomenon in reconstructed phase images is phase discontinuity at the bend contour. Bend contours occur exactly in the region where the Bragg condition is satisfied, and the sign of the excitation error is reversed on the left and right sides, which is understood to be caused by an inversion of the ratio of the excitation of the two Bloch waves. In the following, we will attempt to explain such a phenomenon using mathematical equations.

Phase Modulation of Wedge-Shaped Specimen

According to the dynamical theory of electron diffraction (Bethe method), the wave function ψ_0 of a transmitted electron beam (which corresponds to the object wave in the terminology of electron holography) that propagates a distance t in a crystal specimen is expressed by a linear combination of Bloch waves [31]:

$$\psi_0 = \sum_j \alpha^{(j)} C_0^{(j)} \exp\left(2\pi i \left| \vec{k}^{(j)} \right| t\right)$$

$$= \sum_j \alpha^{(j)} C_0^{(j)} \exp\left(2\pi i\left(\left|\vec{K}\right| + \gamma^{(j)}\right)t\right)$$

$$= \exp\left(2\pi i\left|\vec{K}\right|t\right) \sum_j \alpha^{(j)} C_0^{(j)} \exp\left(2\pi i \gamma^{(j)} t\right). \tag{4.40}$$

When the crystal has a centrosymmetric structure, $\alpha^{(j)} = C_0{}^{(j)}$. The intensity I_0 of the object wave can then be calculated using the equation

$$I_0 = \psi_0^* \psi_0 = \left|\sum_j \alpha^{(j)} C_0^{(j)} \exp\left(2\pi i \gamma^{(j)} t\right)\right|^2. \tag{4.41}$$

The phase shift ϕ is the difference between the phase of an object wave (4.40) and the phase of a reference wave. The phase of the reference wave is expressed by $2\pi|\chi|t$, where χ is the wave vector of an electron in vacuum. Therefore, ϕ can be calculated using the formula

$$\phi = \text{angle}\{\psi_0 \exp(-2\pi i|\chi|t)\}$$

$$= \text{angle}\left[\left\{\exp(2\pi i|K|t) \sum_j \alpha^{(j)} C_0^{(j)} \exp\left(2\pi i \gamma^{(j)} t\right)\right\} \exp(-2\pi i|\chi|t)\right]$$

$$= \text{angle}\left\{\exp(2\pi i(|K| - |\chi|)t) \sum_j \alpha^{(j)} C_0^{(j)} \exp\left(2\pi i \gamma^{(j)} t\right)\right\}$$

$$= \text{angle}\left\{\exp\left(i\sigma \int_0^t \varphi \mathrm{d}z\right) \sum_j \alpha^{(j)} C_0^{(j)} \exp\left(2\pi i \gamma^{(j)} t\right)\right\}, \tag{4.42}$$

where angle(x) is a function that outputs the angle of the complex number x. Comparing (4.42) with (2.15), we note that the effect of dynamical electron diffraction appears in the summation term in (4.42). Parameters α, C_0, and γ depend on the excitation errors s_g (i.e., the deviation parameter, the distance between a reciprocal lattice point and the Ewald sphere) of the excited Bragg reflections (i.e., the tilting angle of the specimen). Hence, the phase shift also depends on the tilting angle. Equations (4.41) and (4.42) enable us to determine the intensity and phase shift of an object wave under the effect of dynamical diffraction as a function of the specimen tilting angle and thickness.

As an example of a calculation using the above equations, consider the case where 220 reflections are excited in a Si single crystal [32]. To simulate the intensity and phase shift of the transmitted beam, we considered nine beams in a 220 systematic row (000, 220, 440, 660, 880, $\overline{220}$, $\overline{440}$, $\overline{660}$, $\overline{880}$) in our calculations. The effects of the absorption of intensity are not considered. We calculated the dynamical electron diffraction using the Bethe method including an external subroutine that yields the atomic form factors for electron diffraction [13]. The value of φ of Si used in our code is 11.5 V [33].

Figure 4.5 shows the results of a calculation of the dynamical theory of electron diffraction. The columns (a–e) show different diffraction conditions. The upper row (i) shows plots of $\psi_0 \exp(-2\pi i|\chi|t)$ on a complex plane with thickness as a parameter (refer to (4.42)). The thickness ranges from 0 to 300 nm. The data from 20 to 80 nm are plotted as red bold lines. To explain Fig. 4.5(i), a blue vector indicating $\psi_0 \exp(-2\pi i|\chi|t)$ at 20 nm thickness is drawn in Fig. 4.5a(i). The length of the vector corresponds to the square root of the intensity I_0, and the angle between the real axis and the vector corresponds to the phase shift ϕ. The middle row (ii) and the lower row (iii) show the intensity and the phase shift, respectively, plotted as functions of thickness t.

In the center column (Fig. 4.5c), the specimen is under an exact 220 Bragg diffraction condition. In this condition, the trajectory of $\psi_0 \exp(-2\pi i|\chi|t)$ intersects the origin of the complex plane (Fig. 4.5c(i)). At the thickness of the intersection, the length of the vector is zero; thus, the corresponding position in the bright-field image shows dark contrast. Notably, the direction of the $\psi_0 \exp(-2\pi i|\chi|t)$ vector is flipped by π across the origin of the complex plane in Fig. 4.5c(i). Therefore, the phase-shift profile (Fig. 4.5c(iii)) shows jumps of π at the dark thickness fringes. In our calculation, these thicknesses are 50.8, 152.0, and 253.2 nm. These thicknesses approach $(1/2 + n)\xi_{220}$ because the diffraction condition is almost a two-beam condition. Here, ξ_{220} is the extinction distance for the 220 reflection and n is an integer ($n \geq 0$). In our calculation, ξ_{220} becomes 108 nm.

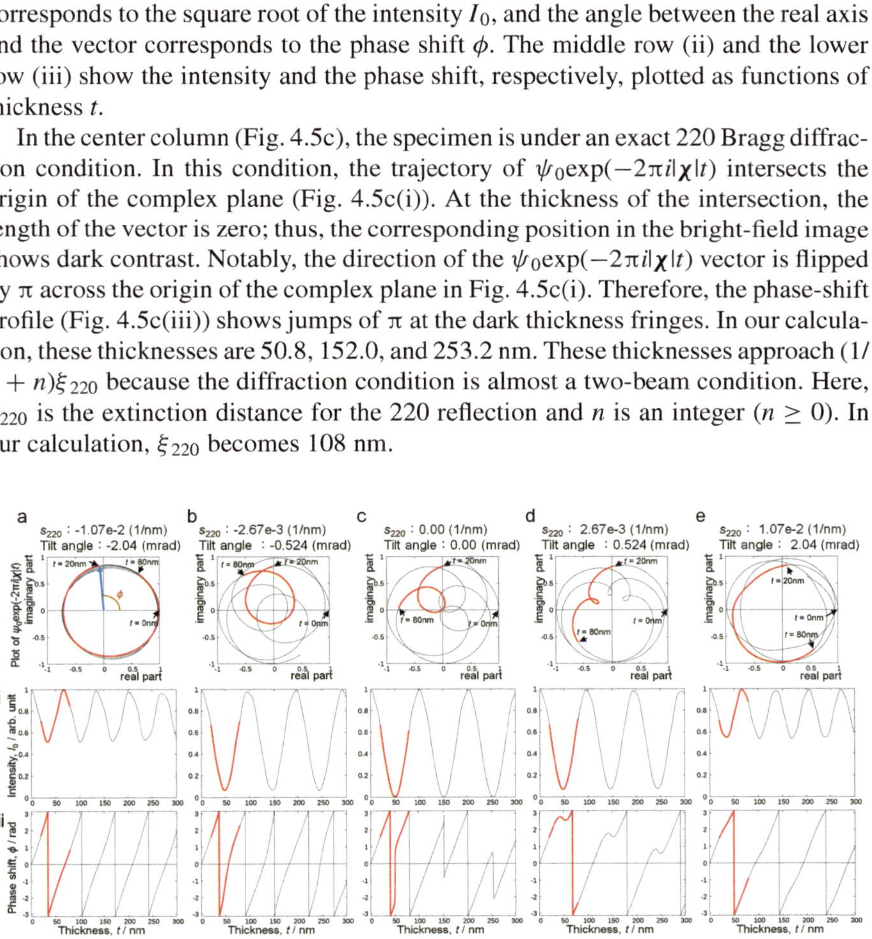

Fig. 4.5 Calculated results based on multi-beam dynamical electron diffraction theory (Bethe method). Columns (a–e) show different diffraction conditions. (i) The complex value $\psi_0 \exp(-2\pi i|\chi|t)$ plotted on a complex plane with thickness as a parameter. (ii) Intensity and (iii) phase shift plotted as a function of thickness. The data from 20 to 80 nm are plotted as red bold lines

The left- and right-side columns in Fig. 4.5 show the results when the sign of the excitation error of the 220 reflection s_{220} is negative and positive, respectively. Figure 4.5(iii) shows that the gradient of the phase shift depends on the sign of the excitation error s_{220} (i.e., when $s_{220} < 0$, the gradient of the phase shift is high, and when $s_{220} > 0$, the gradient of the phase shift is low). This result can be understood by comparing Fig. 4.5b(i) and d(i). In these figures, the trajectory of $\psi_0 \exp(-2\pi i|\chi|t)$, shown as red bold lines for thicknesses from 20 to 80 nm, forms the shape of a small loop. When the sign of s_{220} is negative (Fig. 4.5b(i)), the small loop includes the origin of the complex plane. Thus, the increase of the phase shift is larger than 2π. When s_{220} is positive (Fig. 4.5d(i)), the small loop does not include the origin of the complex plane, which means the increase in the phase shift is less than 2π. Thus, unlike (2.15), (4.42) explains the characteristic phase shift around the Bragg condition.

Figure 4.6 shows bright-field images and reconstructed phase images of a wedge-shaped Si specimen, along with the phase-shift profiles in the region of the specimen indicated by the red rectangles in the bright-field and reconstructed phase images. The specimen shown in Fig. 4.6a is under a non-Bragg condition, and that shown in Fig. 4.6b is under an exact 220 Bragg diffraction condition. Under the non-Bragg condition, no thickness fringes are observed in the bright-field image and the phase shift increases in proportion to the specimen thickness. Under the exact 220 Bragg diffraction condition, thickness fringes appear in the bright-field image and the phase shift jumps by π (indicated by red arrows in the reconstructed phase image and the phase-shift profile), where the dark thickness fringes appear in the corresponding bright-field image. These π jumps cannot be explained by (2.15). However, they can be explained by the dynamical theory of electron diffraction (Fig. 4.5c).

Next, we systematically tilted the direction of the incident electron beam around the 220 Bragg diffraction condition using a beam-tilt function of the illumination system. The bright-field images, the reconstructed phase images, and the phase-shift profiles for each tilt condition are shown in Fig. 4.7a–n. Figure 4.7o and p are diffraction patterns obtained under the diffraction condition corresponding to Fig. 4.7a and n, respectively. From the movement of the crossing point of the Kikuchi lines indicated by red arrows in Fig. 4.7o and p, the total tilting angle from Fig. 4.7a–n is estimated to be 5.04 mrad. The tilting was changed in constant steps (0.39 mrad) such that, for the tilting condition in Fig. 4.7a–g, the excitation error for the 220 reflection s_{220} is negative (which means that the reciprocal lattice point of 220 is located outside the Ewald sphere) and that, for the tilting condition in Fig. 4.7h–n, s_{220} is positive (the reciprocal lattice point of 220 is located inside the Ewald sphere). Notably, the gradient of the phase-shift against the specimen thickness depends on the sign of the excitation error, as discussed in the simulation part (i.e., when $s_{220} < 0$, the gradient of the phase shift is high, and when $s_{220} > 0$, the gradient of the phase shift is low). The complicated phase shifts near the thickness of $(1/2 + n)\xi_{220}$, which was expected on the basis of the simulation, were also observed experimentally.

The complicated phase shifts of an object wave around the Bragg diffraction condition, such as a jump in the phase shift and an effective inner potential depending on

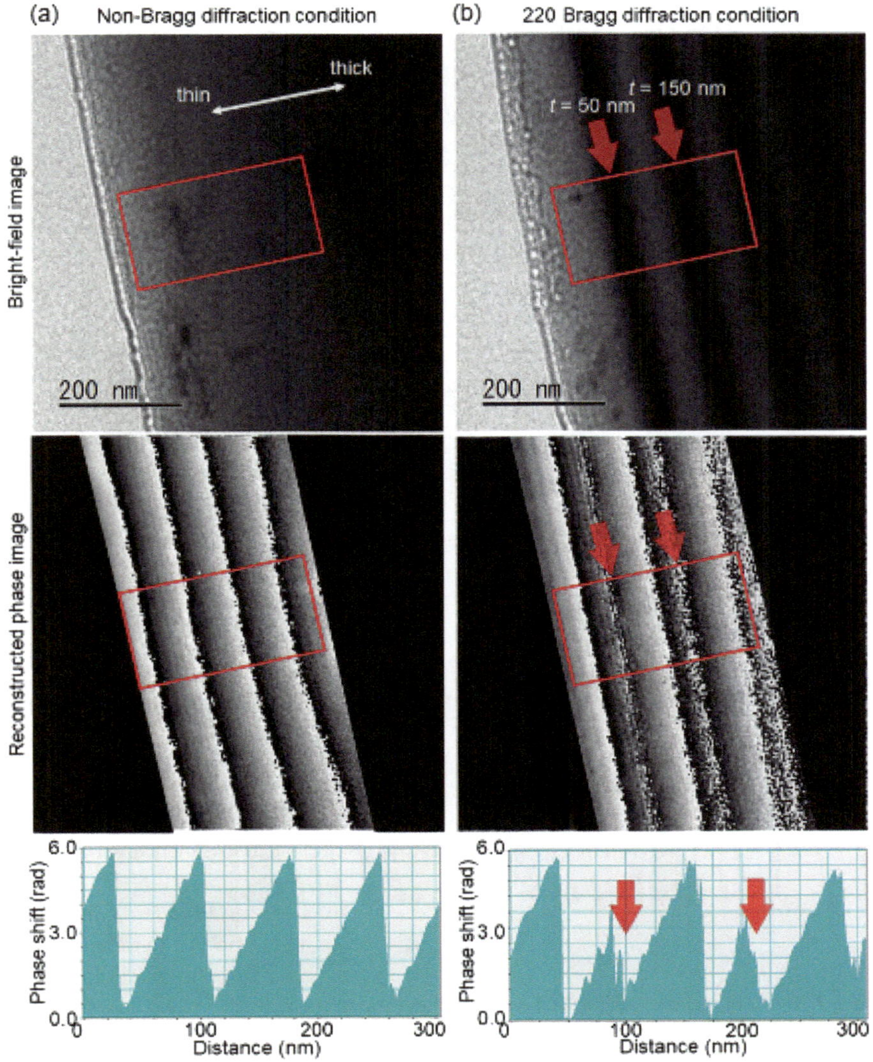

Fig. 4.6 Bright-field images, reconstructed phase images, and phase-shift profiles of wedge-shaped Si crystal. **a** Strong Bragg reflections are not excited. **b** Bragg diffraction condition with 220 reflection is exactly satisfied. The red arrows indicate the positions of dark thickness fringes. The thicknesses of the positions are ~50 and ~150 nm, which were estimated by calculations

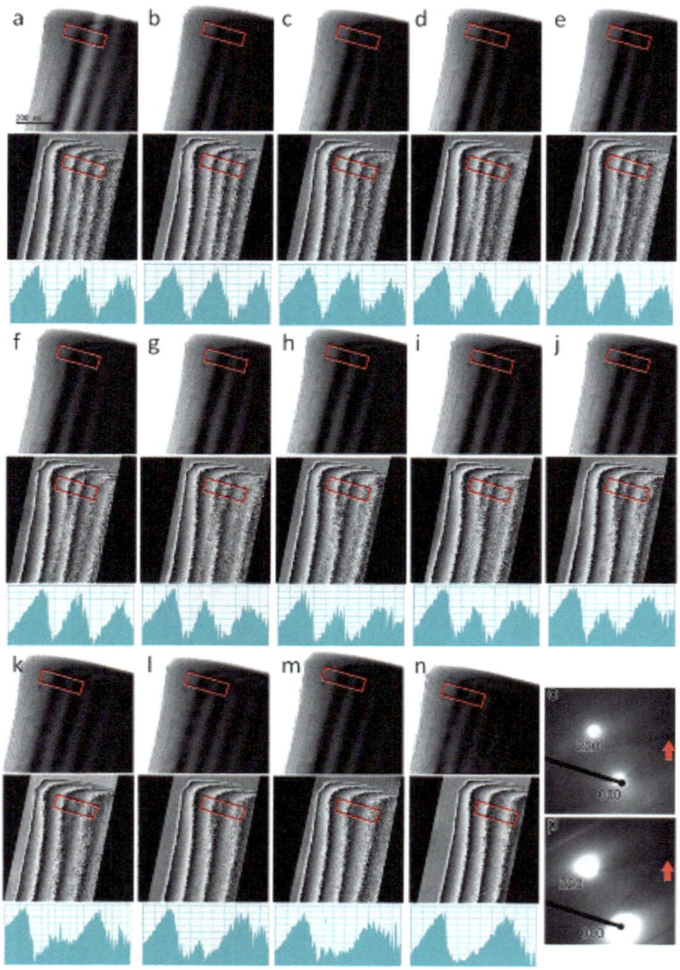

Fig. 4.7 **a–n** Tilt series of bright-field images (upper rows), reconstructed phase images (middle rows), and phase-shift profiles (lower rows, phase shift (rad) versus distance (nm)) for wedge-shaped Si crystal around 220 Bragg diffraction condition. Tilting is changed by a constant step with the excitation error of the 220 reflection s_{220} ranging from negative in **a–g** to positive in **h–n**. **o**, **p** Diffraction patterns obtained under the diffraction conditions of **a** and **n**. The tilt angle between these two diffraction patterns is estimated from the movement of a cross point of Kikuchi lines, which is indicated by a black arrow, to be 5.04 mrad along an axis perpendicular to a reciprocal lattice vector of the 220 reflection

the diffraction conditions, are explained by the dynamical theory of electron diffraction. Near a Bragg condition, many Bloch waves with slightly different wavelengths are excited; the excitation rate of each Bloch wave depends on the incident beam direction. By considering the effect of dynamical electron diffraction, we can interpret the phase-shift map acquired by transmission electron microscopy, e.g., electron holography, DPC-STEM (see Sect. 2.6.3), and electron diffractive imaging [34] quantitatively with high accuracy.

Reconstructed Phase Images of Nanocrystalline Material

Figure 4.8 shows bright-field images of the same area of the specimen containing nanocrystalline Fe-Si-B-P-Cu taken before and after the specimen was flipped [35]. The precipitated Fe-rich body-centered cubic (bcc) crystallites appear as bright or dark spots (depending on the diffraction conditions) with diameters of ~20 nm, clearly separated from each other in the amorphous matrix. For precise inversion of the beam direction with respect to the specimen, the specimen was carefully tilted such that the crystallites that showed strong diffraction (i.e., crystallites that appeared as dark grains in Fig. 4.8a) exhibited similar dark contrast in the new field of view (Fig. 4.8b). Electron holograms were recorded with the same beam incidence angle on the specimen. Figure 4.9 shows one of the holograms used in this analysis. The spacing of the interference fringes was 3.66 nm, and every bcc Fe crystallite with a mean grain size of 20 nm was covered by five or six fringes. The fringe contrast at the vacuum region was approximately 31%.

Figure 4.10 shows the results of amplitude and phase reconstruction from holograms. Figure 4.10a is a reconstructed amplitude image obtained by averaging 19 reconstructed amplitude images; the contrast is similar to the bright-field image in

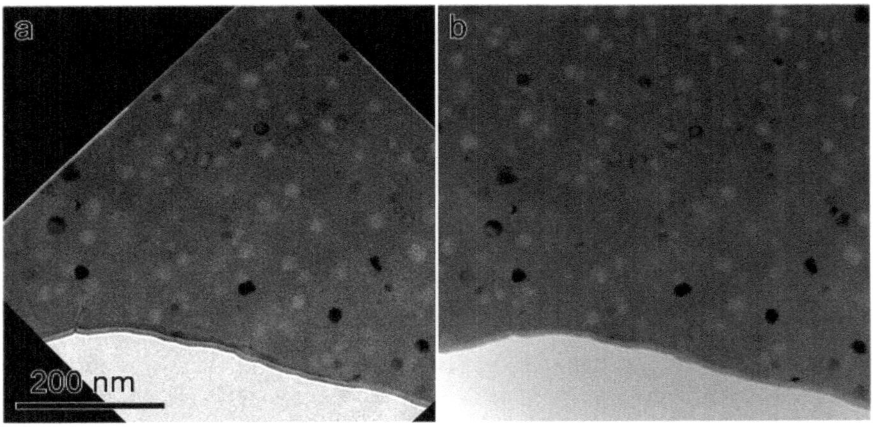

Fig. 4.8 Bright-field images of $Fe_{83.3}Si_4B_8P_4Cu_{0.7}$ specimen acquired from **a** front and **b** back (flipped 180°) sides of specimen. Permission obtained from Elsevier [35]

Fig. 4.9 Representative electron hologram recorded using double-biprism method. Under typical conditions, each crystal grain was covered by 5–6 interference fringes. Permission obtained from Elsevier [35]

Fig. 4.8. Note that the intensity variation observed in the vacuum region (e.g., in Fig. 4.10a above the 200 nm scale bar) is an artifact related to the size of the aperture during amplitude reconstruction using Fourier transformation. Figure 4.10b is a reconstructed phase image obtained by averaging 19 reconstructed phase images. The phase information was amplified by a factor of 10. The contour lines are strongly modulated where they overlap with the nanocrystals. This phase image contains information about both the electric and magnetic fields. Figure 4.10c is an unwrapped phase image because of the electric field extracted by time-reversal operation of an electron beam. The contrast becomes generally brighter toward the inside of the specimen, which indicates that the thickness of the specimen increases toward the inside. Notably, the contrast of some nanocrystallites is brighter or darker than that of the matrix phase. These particles are considered to be in a state where the crystal orientation is close to the Bragg diffraction condition. According to the dynamical theory of electron diffraction, the value of the effective inner potential of the crystal against the incident electron beam changes substantially upon reversal of the sign of the excitation error in the vicinity of the Bragg diffraction condition. Notably, in Sect. 4.1.5.1, the bottom row (iii) in Fig. 4.5 shows the thickness dependence of the phase shift when the 220 reflection is excited in a Si crystal. Figure 4.5a and b correspond to excitation error $s < 0$, Fig. 4.5c corresponds to $s = 0$, and Fig. 4.5d and e correspond to $s > 0$. The slope of the phase shift versus specimen thickness changes dramatically when the sign of s is changed [32]. Thus, nanocrystallites with higher contrast in Fig. 4.10c are considered to have a larger effective inner potential with a negative excitation error, whereas nanocrystallites with darker contrast have a smaller effective inner potential with a positive excitation error. Figure 4.10d shows a phase image due to the magnetic field (i.e., a magnetic induction map). The phase was amplified by a factor of 10. The contour lines, which correspond to lines of magnetic flux, are smoother than in Fig. 4.10b because the effects of the specimen thickness and the diffraction effect have been removed. Note that the magnetic induction is aligned in the direction parallel to the edge of the specimen and that no magnetic domain walls are observed in this field of view. When a soft magnetic material is processed into a thin film such as a specimen for transmission electron microscopy, the in-plane direction generally becomes the easy axis of magnetization because of the shape anisotropy, and the magnetic induction tends to cause a closed domain structure that suppresses magnetic flux leakage to the outside of the specimen in the absence of an external magnetic field. Figure 4.10d shows part of the closure domain because the specimen was fabricated by ion polishing and the thin-film area extends outside the field of view. In Fig. 4.10b–d, the phase fluctuates substantially at the positions indicated by the red arrowheads. These nanocrystallites are likely in the vicinity of the exact Bragg diffraction condition, and the phase change is highly sensitive to the angle of the specimen when the sign of the excitation error is reversed; therefore, extraction of the magnetic field information fails at these locations. If the sign of the excitation error is not reversed when the specimen is reversed in the time-reversal operation of an electron beam, then such an extreme phase change is considered unlikely to occur. For most of the other nanocrystallites, the contour lines pass smoothly.

4.2 Effect of Strong Fields

4.2.1 Electric Field

In the explanation in Sect. 2.3, the hologram is obtained with the object wave and the reference wave, the latter of which corresponds to the vacuum region with a field-free region. In case of a strong field such as an electric field due to the charging effect, obtaining the reference wave in the field-free region is difficult. In such cases, a simulation that takes into account the stray field effect in the reference wave region is carried out. In the following, some experimentally effective methods to suppress the stray electric field in the reference wave region are noted with examples of electric field analyses of toner particles.

Use of Double-Probe Piezodriving Holder

Figure 4.11a shows a schematic of a double-probe piezodriving holder [36]. A micrograph of the top portion of this specimen holder is shown in Fig. 4.11b. The holder has two arms for mounting two metallic probes. The probes can be manipulated independently in three dimensions by the motion of arms 1 and 2, which are driven by micrometers and/or piezoelectric elements. Each probe has three piezoelectric

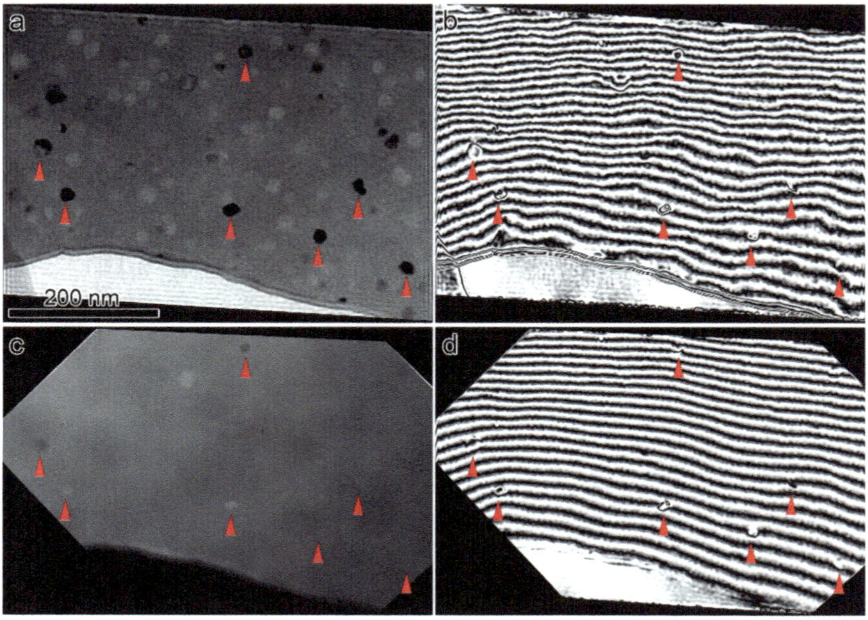

Fig. 4.10 **a** Reconstructed amplitude image obtained by averaging 19 reconstructed amplitude images. **b** Reconstructed phase image obtained by averaging 19 unwrapped reconstructed phase images. The phase information was amplified by a factor of 10. **c** Unwrapped phase image due to electric field. **d** Phase image due to magnetic field (magnetic induction map). The phase was amplified by a factor of 10. Red arrowheads indicate the positions of nanoparticles that appear to be near the Bragg condition. Permission obtained from Elsevier [35]

elements, which enable the fine movements necessary for bringing the probe tip in contact with the specimen. The tips of the needles are sharpened, as shown in Fig. 4.11c, using an FIB system. The probes can be used for various purposes, such as applying a large electric field and/or electric current [37, 38] and producing a large magnetic field with a magnetic needle made of a permanent magnet.

An example of the application of the double-probe piezodriving holder is imaging an electric field produced by mechanical friction, which is an important phenomenon in electric printing technology. The charge on a toner particle, which has yet to be evaluated, can be determined using the present technique. Such information will be useful for producing toner particles with optimized properties. Two technical problems are encountered when using electron holography for such an electric field analysis. The first problem is unwanted charging of the toner particles due to electron irradiation: the emission of secondary electrons makes the toner particles positively charged although mechanical friction by itself induces negative charges. To avoid this problem, a thin-plate Mo shield can be placed over the toner particles to screen them from the electron beam [39]. Although the toner particles cannot be imaged when an Mo shield is used, the charge-induced electric field outside the shield can be observed

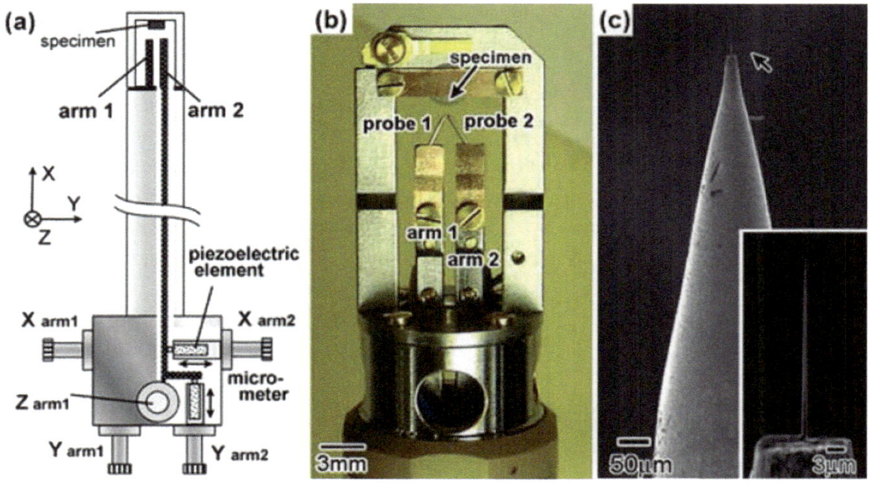

Fig. 4.11 **a** Schematic of double-probe piezodriving holder. When the specimen holder is used, the incident electrons are parallel to the Z-direction. **b** Optical micrograph showing top portion of specimen holder. **c** Scanning ion microscopy image of Pt–Ir needle, tip of which is sharpened using FIB system (the inset shows enlarged image of needle tip)

by holography (Fig. 4.12b). The second problem is perturbation of the reference wave by the long-range field produced by the triboelectric charge. To prevent this electric field and free the reference wave from the perturbation, another shield (a Pt–Ir plate, connected to ground) can be placed within the hologram (Fig. 4.12a). The position of the shields can be accurately controlled using the piezodriving arms of the specimen holder [36].

The left panel in Fig. 4.13a shows a reconstructed phase image obtained using the experimental setup shown in Fig. 4.13b; note that the phase shift is shown in terms of the direct phase ϕ rather than cos ϕ. Many contour lines are observed between the toner particle of interest and the Pt–Ir shield, whereas such lines are not evident on the right of the shield because of the successful shielding of the electric field. Unfortunately, as shown in Fig. 4.13a, there were many positive charges under the Mo shield (refer to the negative slope shown in the right panel, which represents the phase shift measured along line X–Y). These positive charges are observed because the result includes contributions not only from the toner particle of interest (blue area in Fig. 4.13a) but also from carrier particles and/or other toner particles that are out of view.

To eliminate the undesired signal components, we acquired another reconstructed phase image (Fig. 4.13b) for the same field of view without the toner particle of interest using part of the Pt–Ir shield. The true electric field of the toner particle (i.e., the removed toner particle) was obtained by subtracting the phase information shown in Fig. 4.13b from that shown in Fig. 4.13a. The result is shown in Fig. 4.13c. The slope of the phase shift in Fig. 4.13c is opposite those in Fig. 4.13a and b. We

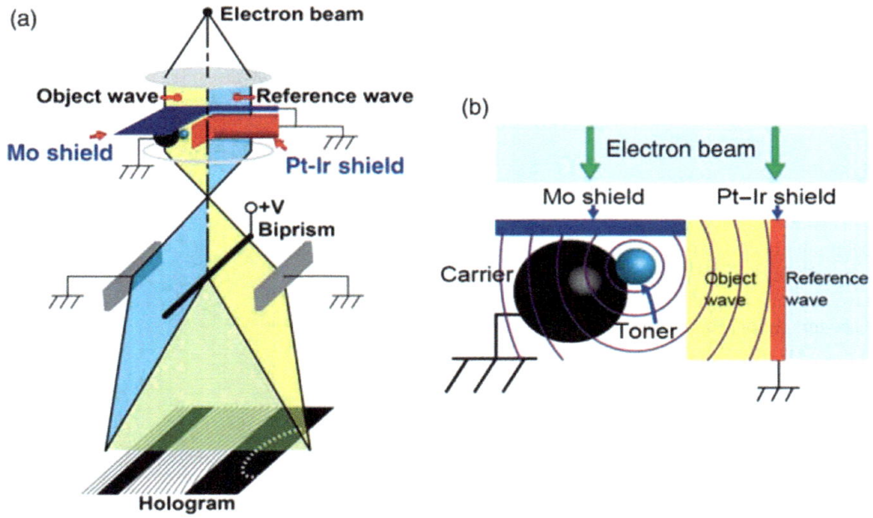

Fig. 4.12 **a** Imaging process in electron holography. **b** Shielding technique for avoiding electron irradiation

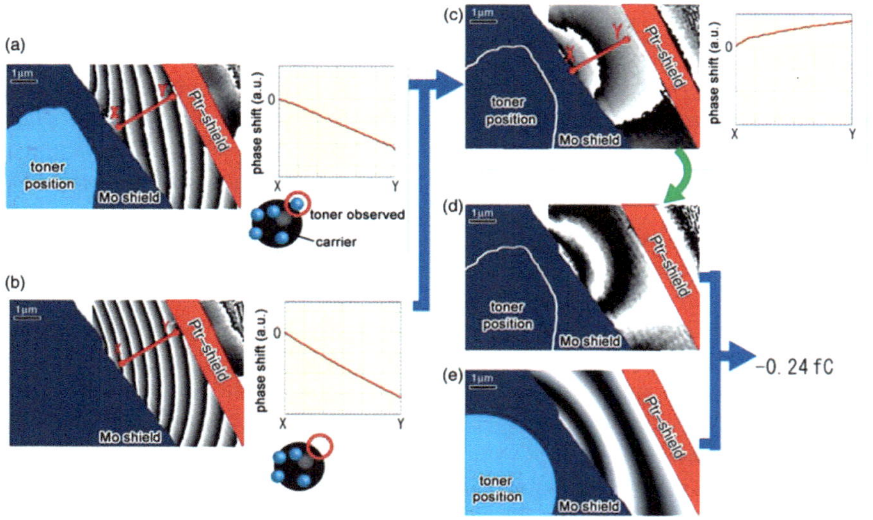

Fig. 4.13 Reconstructed phase images showing triboelectricity of toner particles **a** with toner particle (blue region) beneath Mo shield and **b** after removing toner particle. **c** Result of subtracting phase information of **b** from that of **a**. **d** Result of **c** presented in terms of cosϕ. **e** Result obtained through computer simulation

thus conclude that the toner particle was negatively charged. The result shown in Fig. 4.13c is presented in terms of cos ϕ in Fig. 4.13d.

Figure 4.13e shows simulation results for the equipotential lines around the toner particle, with the phase shift expressed in terms of cos ϕ. When an electric charge of −0.24 fC was assumed, the simulated image was in satisfactory agreement with the observed image in Fig. 4.13d.

Split-Illumination Electron Holography

Split-illumination electron holography can be used to overcome a strong stray field existing near a specimen. As shown in Fig. 4.14a, a biprism was added in the illumination system to split the coherent electron wave into two coherent waves [40, 41]. One of the split waves is used as the object wave illuminating an observation area far from the specimen edge, and the other is used as the reference wave passing through the vacuum outside the specimen. These waves are made to interfere using two biprisms in the imaging system to form a hologram. A high-contrast hologram can be obtained in an area far from the specimen edge without reducing the illuminating electron intensity accordingly. In addition, the reference area can be placed away from the specimen to reduce modulation of the reference wave by stray fields from the specimen. The separation between the two waves is controlled by adjusting the voltage of the biprism in the illumination system.

The fringe spacing s and hologram width W are independently controlled by adjusting the voltages applied to the upper and lower biprisms in the imaging system, as in double-biprism electron interferometry [42]. The filament electrode of the upper biprism is placed at the image plane of the objective lens, whereas that of the lower biprism is placed between the crossover point below the magnifying lenses and the hologram plane (Fig. 4.14a).

For precise phase measurements, further improvements have been achieved using double-biprism split-illumination electron holography without Fresnel fringes (Fig. 4.14b) [41, 43]. The upper and lower condenser biprisms are placed between the second and third condenser lenses. The third lens is used to focus the shadow of the upper condenser biprism filament onto the specimen plane. As a result, the Fresnel fringes produced by the upper condenser biprism are not superimposed on the hologram. The lower biprism is placed in the shadow formed between the paths of the two coherent electron waves separated by the upper condenser biprism. The improved system is capable of imaging using only a single biprism located near the crossover plane.

Split-illumination electron holography can be used with a mask plate to analyze the electrostatic interactions between a toner particle and a carrier particle [41, 44]. Model toner specimens containing polystyrene particles, positive or negative toner particles, and a carrier particle were attached to a Cu needle for observation by electron microscopy. “Positive” or “negative” was determined by the polarity of the triboelectric charges.

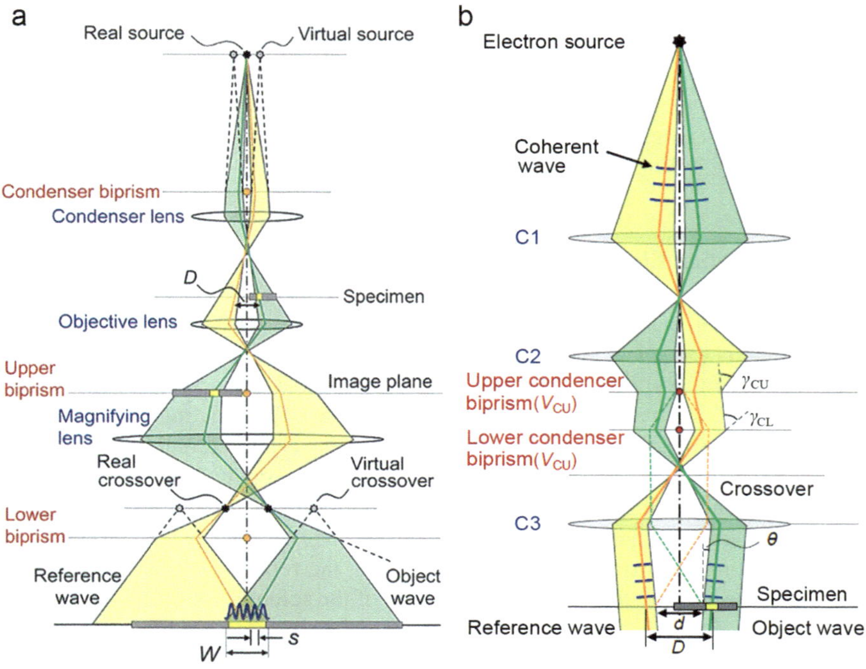

Fig. 4.14 **a** Ray path for (single-biprism) split-illumination electron holography. **b** Optics of illumination system for (double-biprism) split-illumination electron holography. Permission obtained from Elsevier [41]

Figure 4.15a shows the optical system used to analyze the electrostatic potential distribution around a toner particle and the distributions between the toner and carrier particles. To prevent charging of the insulating specimen by electron irradiation, the specimen was placed in a mask plate shadow. The shadow was formed on the specimen plane without Fresnel fringes by adjusting the focusing lens of the mask between the mask and specimen. The mask plate was patterned using an FIB instrument to duplicate the specimen shape. The mask plate was mechanically rotated and moved horizontally. Because the mask was placed in the illumination system, the effects of the secondary electrons emitted from the mask were eliminated through the narrow electron paths in the lens system between the mask and specimen positions.

Split-illumination electron holography can be used to place the reference wave far from the charged specimen to minimize reference-wave perturbations due to specimen charging. The smooth modulations in the reference wave caused by long-range electric fields were simulated in the phase analysis. In the electrostatic potential analysis, a phase image reconstructed from the hologram was compared with the simulated phase image by considering various potential distributions in the specimen. Phase simulations were performed in three dimensions to calculate the phase shifts in the object wave and the reference wave. In these simulations, the electrostatic

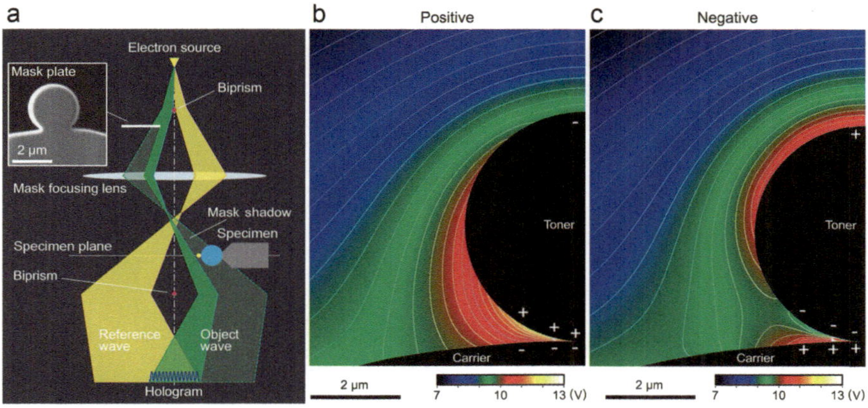

Fig. 4.15 Schematic of optical system for observing and analyzing electric potential distributions of charged insulating specimen. **a** Specimen was placed in shadow of mask plate to prevent charging by electron irradiation. Split-illumination electron holography was used to remove the perturbation of the reference wave caused by the electric field from the charged specimen. **b**, **c** Analyzed electric potential distributions around **b** positive and **c** negative toner particles at specimen plane. Each contour line corresponds to a potential difference of 0.2 V. Permission obtained from Elsevier [41]

potential had axial symmetry along the axis passing through the center of the toner and carrier particles [44].

The analyzed electric potential distributions around the toner particle attached to the carrier particle shown in Fig. 4.15b and c show that the distributions were nonuniform for both the positive and negative specimens. They also indicate that the local charges on the toner particle surface had opposite values for the positive and negative specimens. The charges on the carrier surface facing the toner particle were opposite those on the toner particle surface for both specimen types. The charges on the toner particle surface far from the carrier particle were opposite those on the carrier side of the toner particle.

These findings suggest that dielectrically polarized charges contribute to the electrostatic interactions between toner and carrier particles. The detailed charge distribution on both a toner particle and a carrier particle has yet to be analyzed. The detailed mechanism of the adhesion of charged particles can be understood on the basis of quantitative information about the distribution of charges on toner particles and a carrier particle, as obtained using the proposed technique, resulting in improved electrophotographic printing.

4.2.2 Magnetic Field

In Chaps. 2 and 3, various magnetic domains were observed by Lorentz microscopy and electron holography. Figure 4.16a shows the domain structure of an amorphous

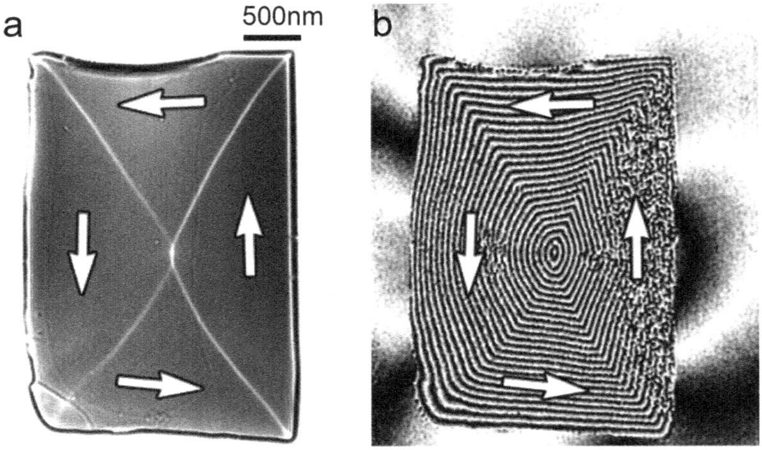

Fig. 4.16 **a** Domain structure of amorphous FeBSi specimen observed by Lorentz microscopy. **b** Reconstructed phase image obtained by electron holography in region corresponding to **a**. Note that a part of the magnetic flux distribution in the magnetic domain in the right part is missing

FeBSi specimen observed by Lorentz microscopy in Fresnel mode. Figure 4.16b is the reconstructed phase image obtained by electron holography in the region corresponding to Fig. 4.16a. The magnetic flux distribution is clearly observed inside the specimen except the right part of the magnetic domain, where the magnetic flux disappears near the specimen edge.

The illustration in Fig. 4.17 shows the ray path for this experimental condition. Figure 4.17a–c correspond to the ray paths with the biprism off. One of the ray paths is indicated in red. Figure 4.17d–f correspond to the condition with the biprism on. In Fig. 4.17a and d, the deflection angle due to the magnetic field of a part of the specimen corresponding to a magnetic domain is small. In Fig. 4.17d, the ray path indicated by red changes to blue, indicating the formation of the object wave under the biprism. However, in Fig. 4.17c and f, the deflection angle is large. In this case, the ray path indicated by red in Fig. 4.17f does not change its color, indicating that the object wave does not form under the biprism; the wave goes into the reference wave region. In the lower part of the hologram in Fig. 4.17f, the missing region of interference fringes is noted. Thus, the corresponding reconstructed phase image does not provide magnetic flux information.

The observed contrast in Fig. 4.16b corresponds to Fig. 4.17e. In Fig. 4.17e, the ray path indicated by red splits into two ray paths indicated by blue and red. The ray path indicated by blue forms the object wave under the biprism, resulting in the reconstructed phase image showing the magnetic flux distribution. However, the ray path indicated by red does not form the object wave and yields no magnetic flux information. The imperfect formation of magnetic flux in a magnetic domain in the right part of the specimen is thus explained by considering such ray paths carefully.

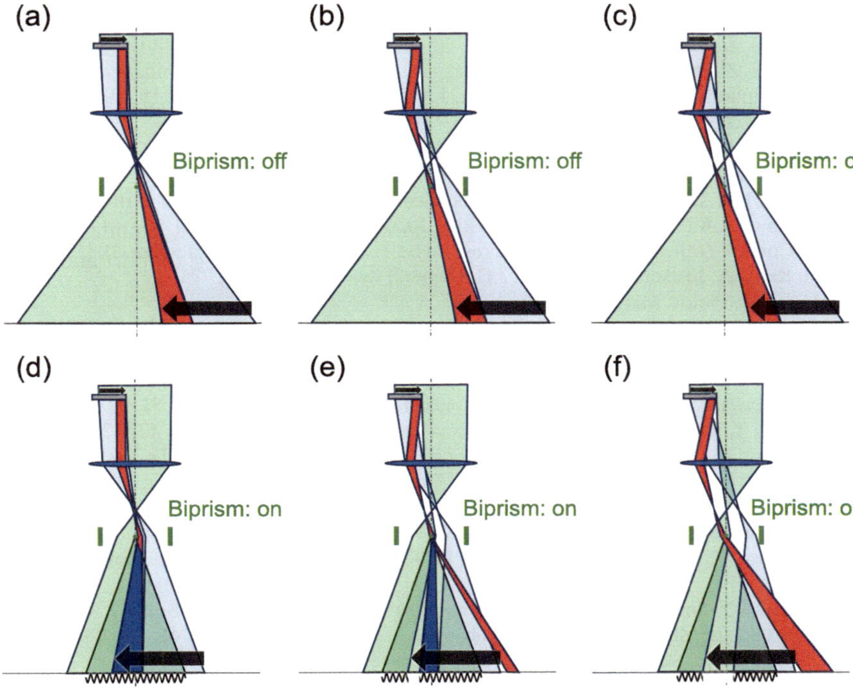

Fig. 4.17 Illustration of ray paths for electron holography for various magnitudes of deflection conditions due to magnetic field of specimen. Ray paths **a**–**c** correspond to the condition of biprism off. Ray paths **d**–**f** correspond to the condition of biprism on. In **a** and **d**, the deflection angle is small, whereas in **c** and **f**, the deflection angle is large. In **b** and **e**, illustrations with a middle deflection angle are shown. Refer to the text for details

Notably, these experimental conditions can be interpreted by assuming the ray path of each incident electron, unlike the treatment of simple plane-wave incidence. This contradiction is related to the so-called wave–particle dualism of the electron. In the next chapter, we will discuss this subject in detail with the spinning linear wave model for the electron.

References

1. C.G. Darwin, Philos. Mag. **27**, 675 (1914)
2. A. Howie, M.J. Whelan, Proc. R Soc. **A263**, 217 (1961)
3. H. Bethe, Ann. Phys. **392**, 55 (1928)
4. C.H. MacGillavry, Physica **7**, 329 (1940)
5. R.D. Heidenreich, J. Appl. Phys. **20**, 993 (1949)
6. N. Kato, J. Phys. Soc. Jpn. **7**, 397 (1952)
7. L. Sturkey, Acta Crystallogr. **10**, 856 (1957)

8. F. Fujimoto, J. Phys. Soc. Jpn. **14**, 1558 (1959)
9. H. Niehrs, Z. Naturf **14a**, 504 (1959)
10. H. Niehrs, Z. Phys. **156**, 466 (1959)
11. C.J. Humphreys, P.B. Hirsch, Philos. Mag. J. Theor. Exp. Appl. Phys. **18**, 115 (1968)
12. G. Radi, Acta Crystallogr. A **26**, 41 (1970)
13. D.M. Bird, Q.A. King, Acta Crystallogr. A **46**, 202 (1990)
14. A. Weickenmeier, H. Kohl, Acta Crystallogr. A **47**, 590 (1991)
15. J.M. Cowley, A.F. Moodie, Proc. Phys. Soc. Sect. B **70**, 486 (1957)
16. J.M. Cowley, A.F. Moodie, Acta Crystallogr. **10**, 609 (n.d.)
17. J.M. Cowley, *Diffraction Physics*, 2nd revised ed. (North-Holland, Amsterdam, 1981)
18. P.M. Kelly, A. Jostsons, R.G. Blake, J.G. Napier, Phys Stat Sol A **31**, 771 (1975)
19. S.M. Allen, Philos. Mag. A **43**, 325 (1981)
20. Y. Tomokiyo, S. Matsumura, T. Eguchi, Proc. XI Th Inter. Congr. Electron Microsc. Kyoto **II**, 1085 (1986)
21. J.M. Zuo, M. Kim, M. O'Keeffe, J.C.H. Spence, Nature **401**, 49 (1999)
22. D. Morikawa, K. Tsuda, Y. Maeda, S. Yamada, T. Arima, J. Phys. Soc. Jpn. **81**, 093602 (2012)
23. Z. Akase, Y. Tomokiyo, Y. Tanaka, M. Watanabe, Phys. C Supercond. **338**, 137 (2000)
24. M. Tanaka, M. Terauchi, *Convergent-Beam Electron Diffraction* (JEOL JTD, Tokyo, 1985)
25. H. Inui, A. Fujii, H. Sakamoto, S. Fujio, K. Tanaka, J. Appl. Crystallogr. **40**, 241 (2007)
26. K. Tsuda, M. Tanaka, Appl. Phys. Express **9**, 071501 (2016)
27. J.C.H. Spence, J. Taftø, J. Microsc. **130**, 147 (1983)
28. D. Shindo, M. Hirabayashi, T. Kawabata, M. Kikuchi, J. Electron Microsc. (Tokyo) **35**, 409 (1986)
29. S. Sakaida, Y. Shimokawa, T. Asaka, S. Honda, Y. Iwamoto, Mater. Res. Bull. **67**, 146 (2015)
30. M. Ohtsuka, S. Muto, K. Tatsumi, Y. Kobayashi, T. Kawata, Microscopy **65**, 127 (2016)
31. C.J. Humphreys, Rep. Prog. Phys. **42**, 1825 (1979)
32. Z. Akase, D. Shindo, Mater. Trans. **60**, 2120 (2019)
33. H.K. Gaukler, R. Schwarzer, Optik **33**, 215 (1971)
34. J. Yamasaki, S. Morishita, Y. Shimaoka, K. Ohta, H. Sasaki, Jpn. J. Appl. Phys. **58**, 120502 (2019)
35. Z. Akase, K. Kimura, T. Saito, K. Niitsu, T. Tanigaki, Y. Iwasaki, P. Sharma, A. Makino, D. Shindo, J. Magn. Magn. Mater. **541**, 168519 (2022)
36. Y. Murakami, N. Kawamoto, D. Shindo, I. Ishikawa, S. Deguchi, K. Yamazaki, M. Inoue, Y. Kondo, K. Suganuma, Appl. Phys. Lett. **88**, 223103 (2006)
37. Y. Murakami, T. Yano, D. Shindo, R. Kainuma, T. Arima, Metall. Mater. Trans. A **38**, 815 (2007)
38. N. Kawamoto, Y. Murakami, D. Shindo, J. Appl. Phys. **107**, 044309 (2010)
39. H. Okada, D. Shindo, J.J. Kim, Y. Murakami, H. Kawase, J. Appl. Phys. **102**, 054908 (2007)
40. T. Tanigaki, Y. Inada, S. Aizawa, T. Suzuki, H.S. Park, T. Matsuda, A. Taniyama, D. Shindo, A. Tonomura, Appl. Phys. Lett. **101**, 043101 (2012)
41. D. Shindo, Z. Akase, Mater. Sci. Eng. R. Rep. **142**, 100564 (2020)
42. K. Harada, A. Tonomura, Y. Togawa, T. Akashi, T. Matsuda, Appl. Phys. Lett. **84**, 3229 (2004)
43. T. Tanigaki, S. Aizawa, H.S. Park, T. Matsuda, K. Harada, D. Shindo, Ultramicroscopy **137**, 7 (2014)
44. T. Tanigaki, K. Sato, Z. Akase, S. Aizawa, H. Soon Park, T. Matsuda, Y. Murakami, D. Shindo, H. Kawase, Appl. Phys. Lett. **104**, 131601 (2014)

Chapter 5
Coherency and Spin

Abstract Interference effect and point-charge behavior of the electron are discussed theoretically on the basis of experimental results obtained using electron holography observation of the electromagnetic field and collective motion of electrons. Taking into account the quantum mechanics and the general theory of relativity, the spinning linear wave (SLW) model of the electron is explained. Diffraction and interference effects of the electron simulated using the SLW model are described. Finally, Larmor precession of the spin magnetic moment of the electron is also interpreted in terms of the SLW model.

In this chapter we discuss the point-charge behavior and interference effect of electrons on the basis of observation of electromagnetic fields by electron holography as well as the relevant quantum mechanics and the general theory of relativity. We also discuss the interference and diffraction effects of the electron on the basis of the spinning linear wave (SLW) model and interpret Larmor precession of the spin magnetic moment with this model.

5.1 Coherency

5.1.1 Coherence Lengths

In this section, the theoretical background and experimental evaluation of coherence length [1] is presented concisely. As shown in Fig. 5.1, the possibility of incident electron interaction in the microscope column is evaluated as follows. For electron holography studies under a high coherence condition, the electron beam current is reduced to 2.0×10^{-11} A at 200 kV. With this low beam current and the velocity

Supplementary Information The online version contains supplementary material available at https://doi.org/10.1007/978-981-97-3537-2_5.

D. Shindo and Z. Akase, *Electron Holography for Electromagnetic Materials*, Springer Series in Materials Science 347, https://doi.org/10.1007/978-981-97-3537-2_5

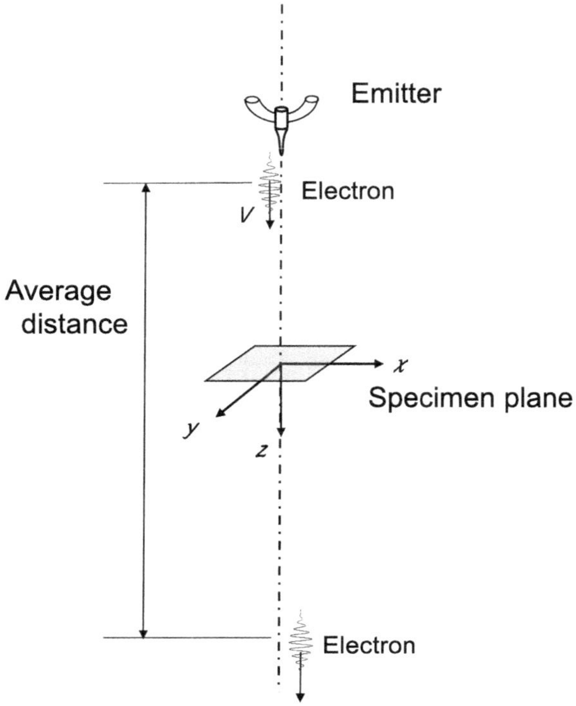

Fig. 5.1 Schematic illustration showing average distance between incident electrons

of 200-keV electrons (2.08×10^8 m/s), the average separation between electrons is estimated to be about 3 m. Since the column of an electron microscope is less than 2 m long, there is on average one electron in the column at a time. Under this condition, the interactions among the incident electrons are considered to be neglected.

The effective coherence length of an electron plane wave [2] has been treated in a manner similar to that used in light optics [3]. Being different from an ideal wave source, i.e., a monochromatic point source, the real source is extended and polychromatic. Thus, the effects of the finite size of the source should be discussed in terms of spatial (or lateral) coherence length L_{SP} given as

$$L_{SP} \simeq \lambda/(2\alpha_i)\,, \tag{5.1}$$

where α_i is the illumination angle as shown in Fig. 5.2. On the other hand, the effects of different energies should be discussed in terms of temporal (or longitudinal) coherence length L_{TE}, as given by

$$L_{TE} \simeq (2E/\Delta E)\lambda, \tag{5.2}$$

where E and ΔE are the kinetic energy and energy spread of the incident electrons, respectively. ΔE can be obtained directly from the energy-loss spectrum. For the use

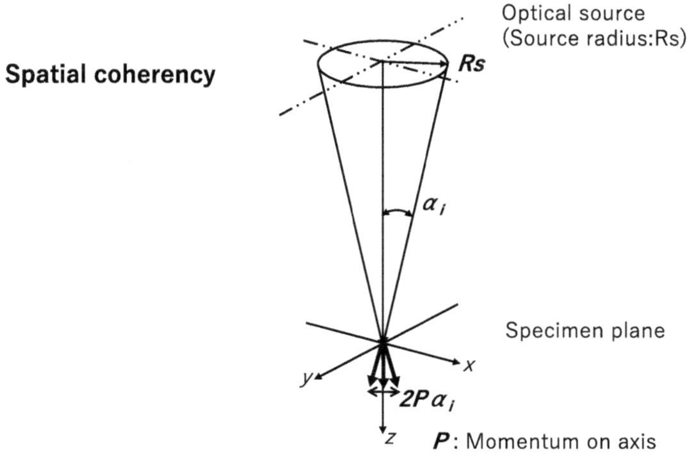

Fig. 5.2 Illumination angle and vertical component of electron momentum relative to optical axis

of a 200-kV electron microscope, $\lambda = 0.0025$ nm, $\alpha_i \simeq 10^{-6}$ rad and $\Delta E \simeq 0.8$ eV. These values result in about 1.3 μm for both L_{SP} and L_{TE}.

Note that L_{SP} is considered to correspond to the uncertainty of the position (Δr) of an incident electron. On the other hand, the uncertainty of the momentum (Δp) is given as

$$\Delta p = \alpha_i \cdot p = \alpha_i h/\lambda. \tag{5.3}$$

From (5.1) and (5.3), the relationship corresponding to the Heisenberg uncertainty principle, i.e.,

$$\Delta r \cdot \Delta p = h/2(\geq h/4\pi) \tag{5.4}$$

is obtained.

5.1.2 Electron Interference and General Relativity

Here, we should note the two features of electrons in the observation of collective motions of electrons in Sect. 3.2.2. As shown in Fig. 5.3 [4], the first is related to secondary electrons whose orbits indicated in red are visualized at the nanometer scale by electron holographic disturbance-free observation (Sect. 1.4.2). The experimental results are consistent with the charge conservation law at the nanometer scale. In other words, the electric field produced by electric charges observed by electron holography corresponds well to Maxwell's equations and is consistent with the special theory of relativity [5].

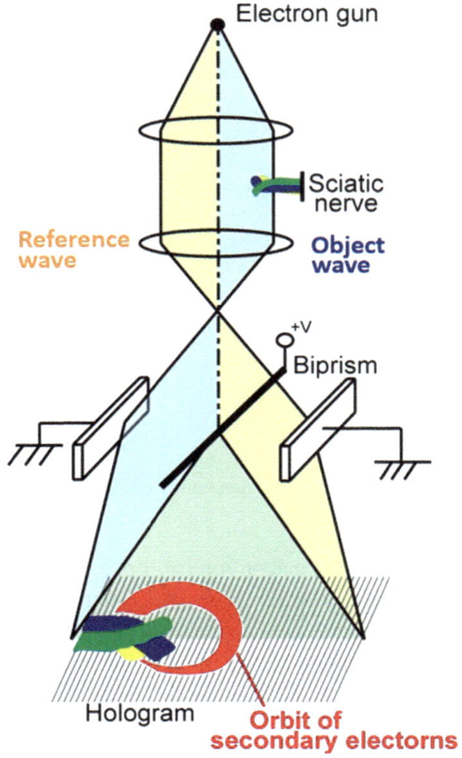

Fig. 5.3 Schematic illustration of experimental setup used to form hologram of microfibrils of sciatic nerve tissue. Permission obtained from Elsevier [3]

The second is related to incident electrons indicated by blue and yellow shading, with which observation is done using interference effects extending to the micrometer scale. The range of interference regions or coherent regions is characterized by the coherence lengths, as discussed above. Note that the coherence lengths depend on the wavelength or momentum, and are basically applicable to any particle, including photons and neutrons, regardless of whether the particle has a charge. Thus, unlike the electric field treated by Maxwell's equations, it was pointed out that the field corresponding to the region where the interference effect occurs cannot be discussed in the framework of the special theory of relativity [4]. The field resulting in the interference effect should be discussed with Einstein's field equations based on the general theory of relativity as discussed in the next section.

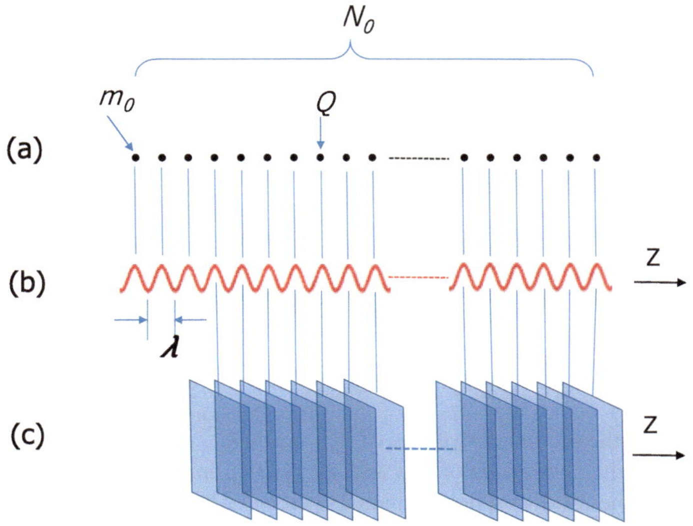

Fig. 5.4 **a** Illustration showing electron consisting of one-dimensional distribution of N_0 point particles used in discussion of gravitational field equations. **b** One-dimensional gravitational field distribution with wavelength λ corresponding to N_0 point particles. **c** Schematic illustration showing plane wave consisting of blue planes with same phase

5.1.3 Infeld and Schild's Approximate Solution to Einstein's Field Equations

About Einstein's field equations in (1.56), Einstein et al. [6] noted that the energy–momentum tensor must be regarded as purely temporary and more or less as a phenomenological entity for representing the structure of matter. They also pointed out that its entry into the equations made it impossible to determine how much the results obtained were independent of the particular assumption made regarding the constitution of matter. According to these indications, they neglected the energy–momentum tensor and assumed that each particle is a singularity of the gravitational field showing that the motions of particle singularities can be determined by the following field equations, i.e.,

$$R_{\mu\nu} - \frac{1}{2} R g_{\mu\nu} = 0. \tag{5.5}$$

Based on these equations, Infeld and Schild [7] used (5.5) in a study of the motion of test particles and discussed the gravitational field corresponding to N_0 point particles. They assumed that the gravitational field dependence on parameter m_0 corresponds to a mass with $m_0 = 1/N_0$ as shown in Fig. 5.4a. They also assumed neutral matter which is applicable to the following discussion of the electron field since the wave field of an electron depends on its mass or momentum and not its charge. By expanding $g_{\mu\nu}$ into a series in terms of m_0:

$$g_{\mu\nu} = \eta_{\mu\nu} + a_{\mu\nu} + (m_0)b_{\mu\nu} + (m_0)^2 c_{\mu\nu} + \cdots, \tag{5.6}$$

where $\eta_{\mu\nu}$ is the Minkowskian metric (see (1.18)), they inserted $g_{\mu\nu}$ in (5.6) into (5.5). In the limit of $m_0 \to 0$ ($N_0 \to \infty$), they obtained the following field equations

$$B_{\mu\nu} = F_{\mu\nu} + L_{\mu\nu} = 0 \tag{5.7}$$

in terms of $\beta_{\mu\nu}$, which is given in the introduction of equivalent relations, i.e.,

$$\beta_{\mu\nu} = b_{\mu\nu} - \frac{1}{2}\eta_{\mu\nu}\eta^{\rho\sigma}b_{\rho\sigma}, b_{\mu\nu} = \beta_{\mu\nu} - \frac{1}{2}\eta_{\mu\nu}\eta^{\rho\sigma}\beta_{\rho\sigma}. \tag{5.8}$$

$F_{\mu\nu}$ involves the second derivatives of $\beta_{\mu\nu}$, while $L_{\mu\nu}$ contains nonlinear terms of the products of the derivatives of $\beta_{\mu\nu}$ and $a_{\mu\nu}$. In the limit of $m_0 \to 0$ around point Q in Fig. 5.4a, the following field equations can be obtained

$$F_{\mu\nu} = 0. \tag{5.9}$$

Eventually, from (5.9), the following linear equations can be obtained,

$$\eta^{\rho\sigma}\frac{\partial}{\partial x^\rho}\frac{\partial}{\partial x^\sigma}\beta_{\mu\nu} = 0, \tag{5.10}$$

$$\eta^{\rho\sigma}\frac{\partial}{\partial x^\sigma}\beta_{\mu\rho} = 0 \tag{5.11}$$

($\sigma = 0, 1, 2, 3; x^0 = ct, x^1 = x, x^2 = y, x^3 = z$).

It should be note that (5.10) and (5.11) have similar structures to those of Maxwell's equations for electromagnetic potentials in (1.24) and (1.26) without a charge and current. Thus, the solutions to these equations are expected to correspond to a waveform similar to that of an electromagnetic wave.

5.1.4 Spinning Linear Wave (SLW) Model

Based on the results obtained by Infeld and Schild noted in the above section, the SLW model could be proposed [1] as outlined below. First, the finite conditions for the electron; N_0 is a large number but not infinity; i.e., $N_0 \to N_e$, and thus $m_0 \to m_e$ with the relationship

$$m_e \cdot N_e = 1. \tag{5.12}$$

On these conditions, the form of the matter wave given by the last term in (1.55) is assumed to be a solution to these equations for the electron. In the derivations of (5.10) and (5.11), the electron is considered to consist of a one-dimensional distribution of particles shown in Fig. 5.4a, thus its gravitational field is also linear in the z-direction. The corresponding wave function is assumed to be given in a one-dimensional form, i.e.,

$$\psi_{LW}(z,t) = \exp[i(kz - \omega t)]\,. \tag{5.13}$$

This one-dimensional wave function satisfies the Schrödinger equation. For a fixed t (= 0) of the linear wave ψ_{LW}, cos(kz) is shown in Fig. 5.4b comparing with the regularly distributed wavefronts of the plane wave ψ_{PW} as illustrated with blue planes in (c). Each linear wave was assumed to rotate or spin quickly around the particle position due to the quantized spin angular momentum with a magnitude of $h/(4\pi)$, called the SLW (spinning linear wave) [1].

The linear wave whose length is given by

$$L_{LW} = N_e \cdot \lambda \tag{5.14}$$

is schematically illustrated in Fig. 5.5a where the incident direction of the electron (SLW) is parallel to the z-direction as indicated by the thick red arrow, while the spinning motion of the linear wave is represented by the thin red arrows.

The length of a linear wave (L_{LW}) is considered to exceed the effective coherence lengths of the spatial coherence length (L_{SP}) and the temporal coherence length (L_{TE}) shown in Fig. 5.5b. Since the effective coherence lengths at the micrometer scale are much greater than the length of the interatomic distance (L_{ID}) at the nanometer scale, we have

$$L_{LW} > L_{SP}, L_{TE} \gg L_{ID}, \tag{5.15}$$

L_{LW} is much greater than the scale of the interatomic distance (L_{ID}), and thus the path traced by the SLW on the specimen is a straight line as indicated with a red line l_1 in Fig. 5.5a.

5.1.5 Quantum Phenomena Formulated with SLW Model

Based on the geometrical configuration of a transmission electron microscope as illustrated in Fig. 5.6, we outline the formulations of diffraction patterns and interference fringes [1].

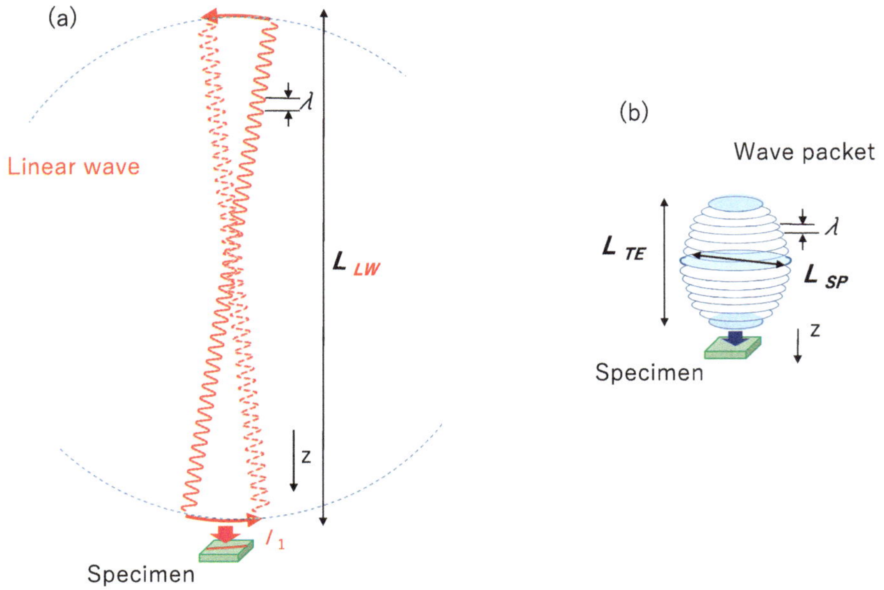

Fig. 5.5 **a** Illustration showing motion of linear wave with length L_{LW}. Spinning motion is indicated by thin red arrows; direction of electron incidence parallel to z-direction is indicated by thick red arrow. **b** Illustration of wave packet moving in z-direction; sizes of spatial and temporal coherence lengths are L_{SP} and L_{TE}, respectively

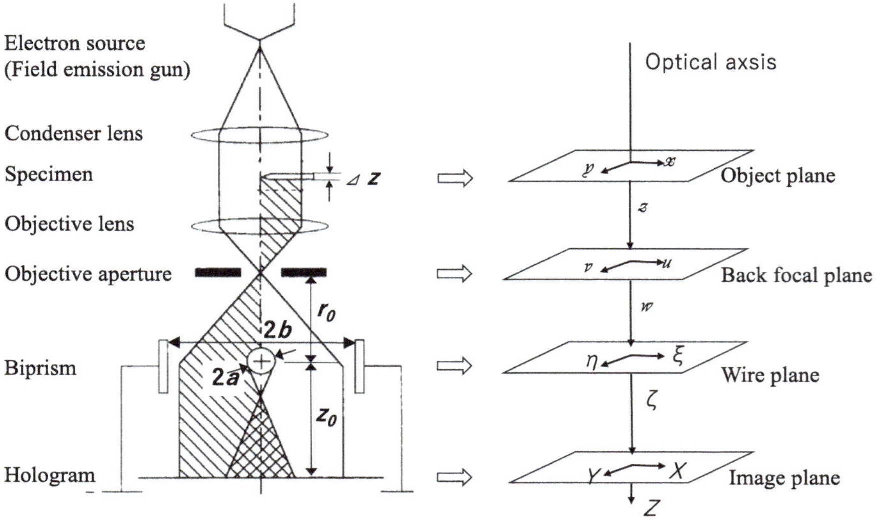

Fig. 5.6 Geometrical configuration of transmission electron microscope with corresponding coordinate systems on right. Shaded region indicates object wave

Diffraction Patterns

As illustrated in Fig. 5.6, with a thin material with thickness Δz in the object plane, we discuss the transmission function with a simple phase object approximation [8, 9], i.e., $\mu = 0$ in (4.32):

$$q(x, y) = \exp\left[i\sigma\varphi(x, y)\Delta z\right] \tag{5.16}$$

When the specimen is homogeneous, the integration of the phase shift due to the scalar potential in the z-direction is replaced with the projected two-dimensional scalar potential distribution for specimen thickness Δz; i.e.,

$$\sigma \int_0^{\Delta z} \varphi(x, y, z)dz = \sigma \cdot \varphi(x, y) \cdot \Delta z, \tag{5.17}$$

where σ is the interaction constant given by (2.16).

The scattering amplitude is given by

$$\psi(u, v, t) = C\exp(-i\omega t) \iint q(x, y)\exp(-2\pi i(ux + vy))dxdy, \tag{5.18}$$

where C is a constant, and u, v are the coordinates of the back focal plane shown in Fig. 5.6. The double integrals correspond to the two-dimensional Fourier transformation [8, 9]. Thus, the relative diffraction intensity distribution is independent of time t and is simply given as

$$\psi^*(u, v)\psi(u, v) = \left| \iint q(x, y)\exp(-2\pi i(ux + vy))dxdy \right|^2. \tag{5.19}$$

In Fig. 5.7a, the plane wave incidence is shown and the motion of an incident SLW is illustrated in (b). In both figures, the thick arrows correspond to the directions of the incident and diffracted electrons, while the thin arrows indicate the translational shift due to the spinning motion of the linear wave in (b).

Under the Bragg condition, i.e., $2d\sin\theta = n\lambda$, the phases of the incident plane wave at positions p_1 and p_2 and also those of the diffracted plane wave at p_3 and p_4 are the same as shown in Fig. 5.7a. On the other hand, when a linear wave moves from the left to the right, the phases of the linear wave at q_1 and q_2 are, in general, different due to its movement parallel to the thick arrows. Under the Bragg condition, the phases of the diffracted linear wave at q_3 and q_4 are different, but the path difference between the linear wave at the left (q_1–q_3) and right (q_2–q_4) positions is $n\lambda$, resulting in the maintenance of its waveform. Note that the maintenance of the waveform of the SLW depends on the Bragg diffraction angle θ, the interatomic distance d and wavelength λ, but not on the phase change due its movement.

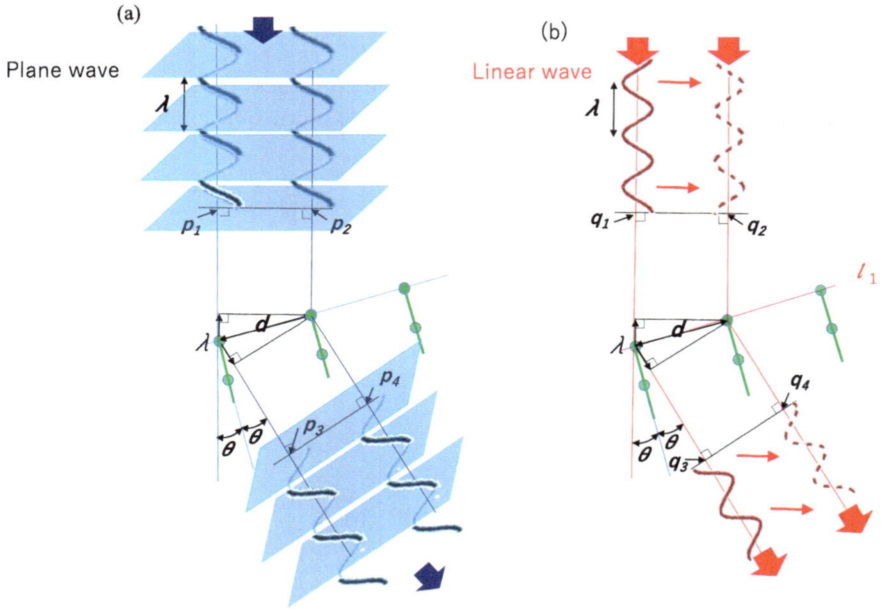

Fig. 5.7 **a** Illustration showing geometrical configuration of incident and diffracted plane waves under Bragg condition in object plane. Directions of waves are indicated by thick blue arrows. Green circles correspond to atoms. **b** Illustration showing diffraction process of spinning linear wave in object plane. Directions of incident and diffracted waves are indicated by thick red arrows. Thin red arrows indicate direction of translational motion of linear wave due to its spinning motion

Since the maintenance of the waveform of the SLW is independent of the phases at p_1 and p_2, the degree of its waveform maintenance along the line trace l_1 in Fig. 5.5a can be simply evaluated as the magnitude of the interference effect when the phases of the SLW along this line are assumed to be equal. Thus, the scattering amplitude is evaluated using (5.18), assuming the area for the two-dimensional integral to be a narrow band region, i.e., l_1 in Figs. 5.5a and 5.7b, i.e.,

$$\psi_1(u, v, t) = C\exp[-i\omega t_1] \iint_{l_1} q(x, y)\exp(-2\pi i(ux + vy))dxdy, \tag{5.20}$$

where $\iint_{l_1} \ldots$ is the domain of integration area l_1. The scattering amplitude for an electron for the hth incidence $\psi_h(u, v, t)$ is accordingly given as

$$\psi_h(u, v, t) = C\exp[-i\omega t_h] \iint_{l_h} q(x, y)\exp(-2\pi i(ux + vy))dxdy. \tag{5.21}$$

The relative diffraction intensity distribution averaged for n incident electrons is thus given as

$$I(u,v) = \frac{1}{n}\sum_{h=1}^{n}\psi_h^*(u,v,t).\sum_{k=1}^{n}\psi_k(u,v,t)$$

$$= \frac{1}{n}\sum_{h=1}^{n}\overline{\left\{\exp(-i\omega t_h)\iint_{l_h} q(x,y\exp\left[-2\pi i(ux+vy)\right]dxdy\right\}}$$

$$\sum_{k=1}^{n}\left\{\exp(-i\omega t_k)\iint_{l_k} q(x,y)\exp\left[-2\pi i(ux+vy)\right]dxdy\right\} \quad (5.22)$$

The overbar $\overline{\{\cdots\}}$ corresponds to the conjugate complex value. The positions and directions of lines l_1, l_2, ... l_n are basically random and the phases of $\omega\, t_1$, $\omega\, t_2$, ... $\omega\, t_\mathrm{n}$ are uncorrelated, and thus, the summation of the product

$$\overline{\left\{\exp(-i\omega t_h)\iint_{l_h} q(x,y\exp\left[-2\pi i(ux+vy)\right]dxdy\right\}}\cdot\left\{\exp(-i\omega t_k)\iint_{l_k} q(x,y)\exp\left[-2\pi i(ux+vy)\right]dxdy\right\} \quad (5.23)$$

will vanish except the summation for $h = k$. Thus, we finally obtain

$$I(u,v) = \frac{1}{n}\sum_{h=1}^{n}\overline{\left\{\iint_{l_h} q(x,y\exp\left[-2\pi i(ux+vy)\right]dxdy\right\}}.$$

$$\left\{\iint_{l_h} q(x,y)\exp\left[-2\pi i(ux+vy)\right]dxdy\right\}$$

$$\approx\left|\iint_{S} q(x,y)\exp\left[-2\pi i(ux+vy)\right]dxdy\right|^2. \quad (5.24)$$

The last line of (5.24) is obtained taking into account $\sum_{h=1}^{n}\iint_{l_h}\ldots$ to be equal to $\iint_S\ldots$ for a large number n: $\iint_S\ldots$ is the domain of integration corresponding to the area of the two-dimensional specimen surface. Thus, the equation shows that the formulation of the intensity distribution for electron diffraction using SLWs is equivalent to the formulation in (5.19), when the number of incident electrons is sufficiently large.

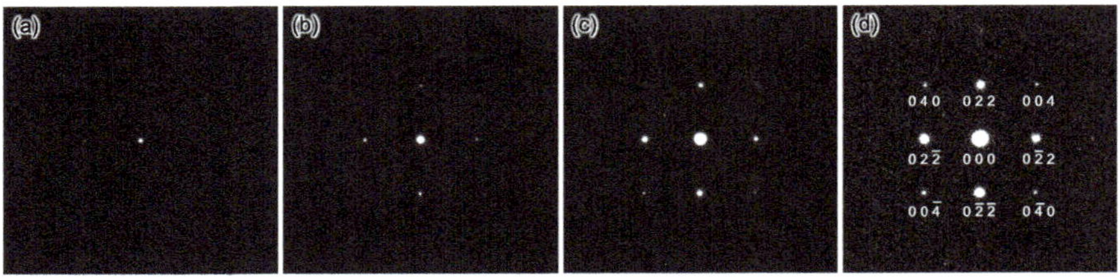

Fig. 5.8 Electron diffraction intensity of Si with [100] incidence as a function of number of incident electrons: **a** 64e, **b** 256e, **c** 1024e, and **d** 4096e. Diffraction patterns of specimen thickness 20 nm were observed with 300-keV electrons

Figure 5.8 shows electron diffraction patterns of Si with [100] incidence obtained with a CMOS camera. As the number of incident electrons increases, the intensity profile of the Bragg diffraction spots gradually forms sharp peaks.

A video of the change in the intensity profiles of these diffraction patterns as a function of the number of incident electrons is presented in Fig. 5.9. To compare these observed intensity profiles as a function of the number of incident electrons, we simulated the intensity profiles on the basis of the SLW model. The trace of an SLW forms a line shape, and the scattering amplitude of this line forms a streak on each Bragg diffraction spot, as shown in Fig. 5.10a.

The relationship between the line shape in the real space and the streak in the reciprocal space was observed in the properties of Fourier transformation described in Sect. 2.2. The streaks for each Bragg spot for n = 2 and 20 incident electrons are shown in Fig. 5.10b and c. When n is large such as 20,000, a sharp intensity peak appears at each diffraction spot, as shown in Fig. 5.10d.

The intensity profile of the electron diffraction of Si simulated using the SLW model is compared with the observed electron diffraction in Fig. 5.11a. The line

Fig. 5.9 Video: Change in electron diffraction intensity profile with increase in number of incident electrons. Detection count in video is converted into number of electrons by dividing scale of vertical axis (detection count) by 50. Video was obtained in collaboration with Dr. Oikawa and Mr. Hashiguchi

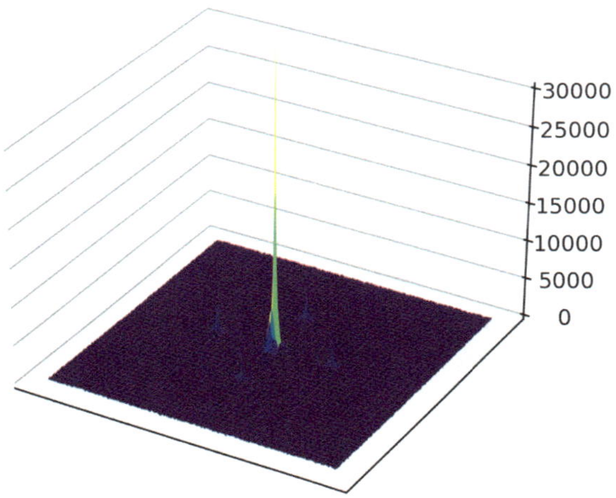

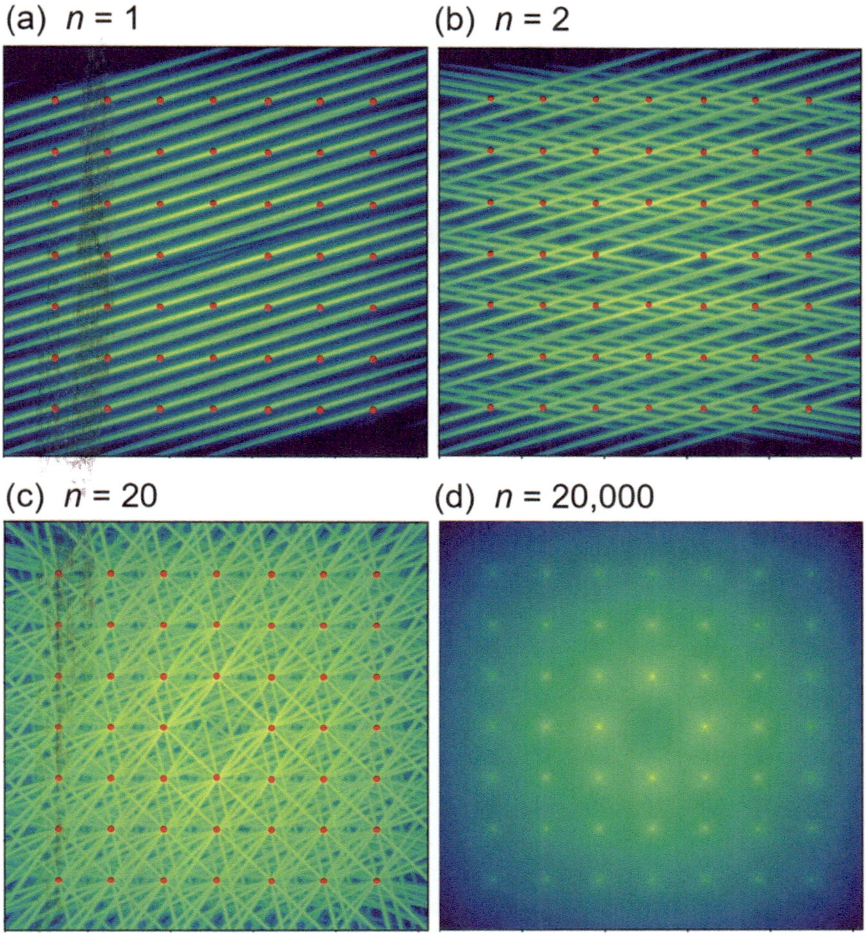

Fig. 5.10 Simulated scattering amplitude at each diffraction spot (indicated by red dots): **a** n = 1, **b** n = 2, **c** n = 20, **d** n = 20,000. Diffraction patterns was simulated for specimen thickness 20 nm and 200-keV electrons

traces of SLWs for $n = 20$ incident electrons striking Si are plotted in Fig. 5.11b. The contribution of each electron to the diffraction pattern was evaluated in accordance with the first line of (5.24) using random numbers to estimate the probability of finding each electron on each line trace. The intensity profile of the simulated electron diffraction gradually increased, resulting in sharp diffraction spots showing good agreement with the observation.

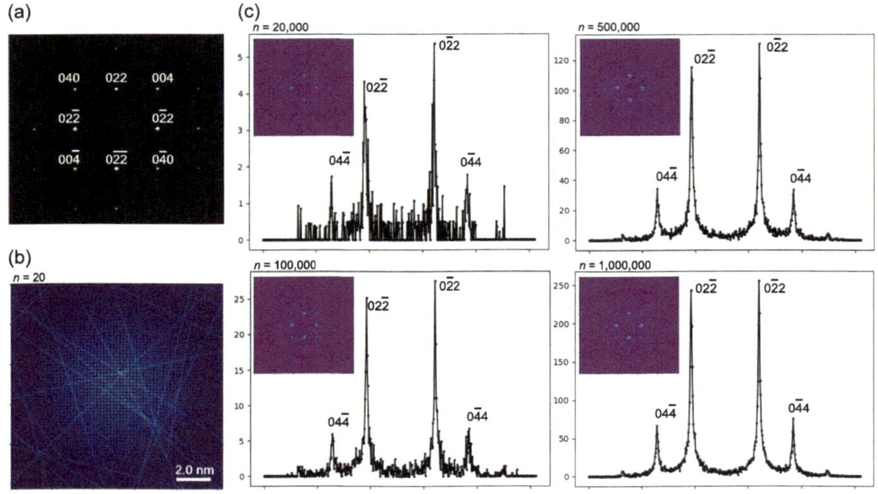

Fig. 5.11 **a** Electron diffraction pattern of Si observed with [100] incidence. **b** Line traces of SLWs for n = 20. **c** Intensity profiles of diffraction spots simulated for Si with [100] incidence for n = 20,000, 100,000, 500,000, 1,000,000. Observation and simulation of diffraction patterns correspond to 20 nm thickness and 200-keV electrons

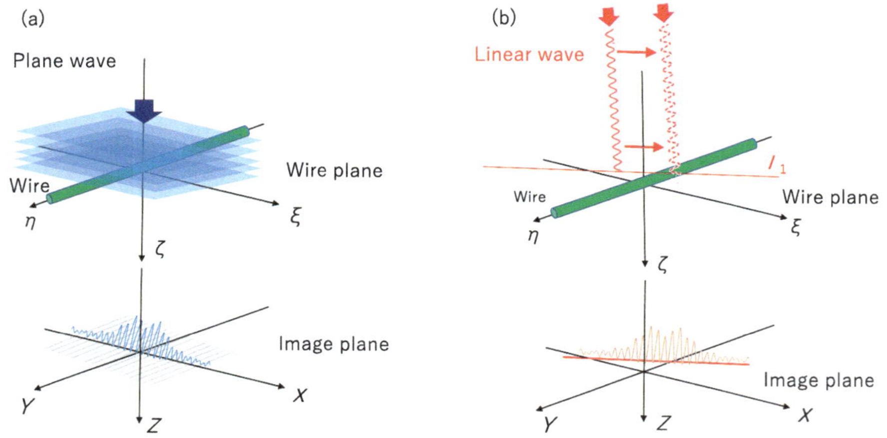

Fig. 5.12 Illustration showing geometrical configuration of incident electron beam and biprism wire. **a** Two-dimensional wave incidence. Formation of interference fringes is illustrated at bottom. **b** Spinning linear wave crosses wire plane perpendicularly. Expected intensity profile of part of the interference fringes is shown at bottom

Electron Interference Fringes

The formation of electron interference fringes is also explained with the SLW model taking into account the degree of its waveform maintenance [1]. The essence of the explanation is as follows.

In Fig. 5.12a, the incidence of the two-dimensional incident wave around a wire plane and the formation of interference fringes in the image plane schematically shown in the upper part and the lower part, respectively. On the other hand, the motion of a SLW is illustrated in the upper part of Fig. 5.12b with the translational shift indicated by the thin red arrows, while the expected intensity profile of the interference fringes for the SLW is shown in the lower. Based on the formulation given by Komrska [10] in Sect. 2.5.1, the amplitude of the interference fringes for the two-dimensional incident wave is given by

$$\psi(X, Y, t) = D\exp(-i\omega t) \int_{-\eta_{\max}}^{\eta_{\max}} \left[\int_{-\xi_{\max}}^{-a} + \int_{a}^{\xi_{\max}} \right] \exp\{i[\phi(\xi, \eta) + \phi_U(\xi, \eta)]\} d\xi d\eta. \tag{5.25}$$

The relative intensity distribution of the interference fringes with the two-dimensional incident wave is also given as

$$\psi^*(X, Y)\psi(X, Y) = \left| \int_{-\eta_{\max}}^{\eta_{\max}} \left[\int_{-\xi_{\max}}^{-a} + \int_{a}^{\xi_{\max}} \right] \exp\{i[\phi(\xi, \eta) + \phi_U(\xi, \eta)]\} d\xi d\eta \right|^2. \tag{5.26}$$

As shown in Fig. 5.13a, the interference effect is mainly produced by the stable phases in two stationary regions in *p'* and *q'* on the left and right sides of the biprism wire along the ξ-axis, respectively. The positions of the stationary regions around the wire plane depend on the positions in the image plane indicated by colored lines. The thick red regions in (b) indicate the main contribution of the two stationary regions in the wire plane to position $X = 0.5$ μm in the image plane. When the phases of the two red lines at $X = 0.5$ μm are equal, a strong interference effect is produced.

When a SLW passes through the wire plane from *p'* to *q'* and maintains its waveform, there is a high probability of electron propagation to position $X = 0.5$ μm in the image plane. Thus, the degree of waveform maintenance for a SLW is assumed to be equal to the magnitude of the interference effect of the two-dimensional incident wave in the region of the line trace (l_1) along the translational motion of the SLW.

Thus, the scattering amplitude of the SLW for the *k*th incidence is given by

$$\psi_k(X, Y, t) = D\exp(-i\omega t_k) \iint_{l_k} \exp\{i[\phi(\xi, \eta) + \phi_U(\xi, \eta)]\} d\xi d\eta, \tag{5.27}$$

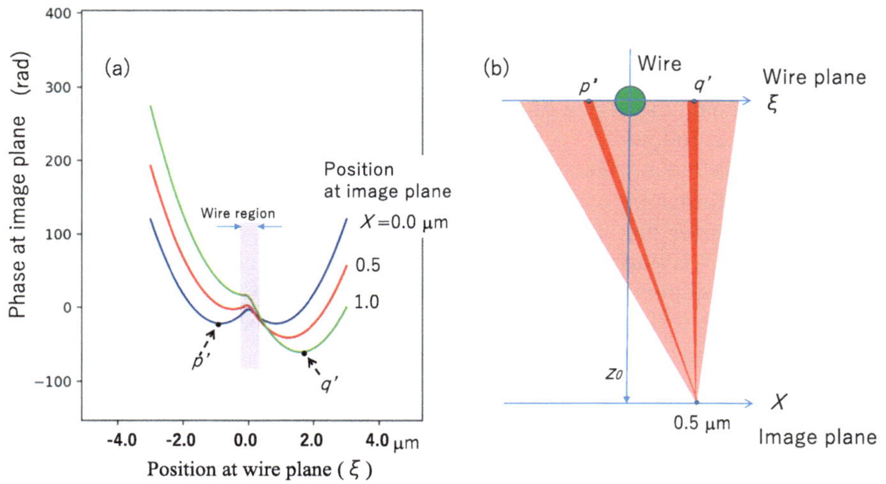

Fig. 5.13 Principle of interference fringe formation with two-dimensional incident wave. **a** Phase change of wave as function of wire plane position along ξ-axis. Colored lines correspond to different positions in image plane (hologram). **b** Enlarged illustration of relationships between stationary regions indicated by thick red lines and point on image plane corresponding to $X = 0.5$ μm. Wire voltage 26 V is assumed

where $\iint_{l_k} \ldots$ is the domain of integration corresponding to the area of line trace l_k on the wire plane except for the wire region. The relative intensity distribution on the X–Y image plane averaged for n electrons is given as

$$I(X, Y) = \frac{1}{n} \sum_{h=1}^{n} \psi_h^*(X, Y, t) \cdot \sum_{k=1}^{n} \psi_k(X, Y, t). \tag{5.28}$$

By neglecting the summation of the products except $h = k$ and setting n to an infinitely large number, (5.28) is given as

$$\begin{aligned} I(X, Y) &= \frac{1}{n} \sum_{k=1}^{n} \overline{\iint_{l_k} \exp\{i[(\xi, \eta) + {}_U(\xi, \eta)]\} d\xi d\eta} \\ &\quad \cdot \iint_{l_k} \exp\{i[(\xi, \eta) + {}_U(\xi, \eta)]\} d\xi d\eta \\ &\approx \left| \iint_{S_w} \exp\{i[(\xi, \eta) + {}_U(\xi, \eta)]\} d\xi d\eta \right|^2, \end{aligned} \tag{5.29}$$

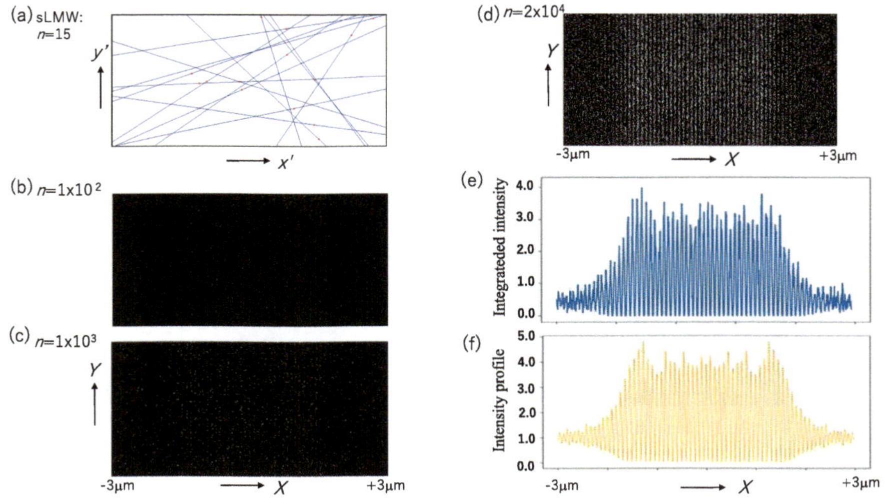

Fig. 5.14 Comparison of observed and simulated intensity profiles of interference fringes. **a** Line traces of 15 SLWs on wire plane (small red dots represent positions of electrons on image plane). **b–d** Simulated interference fringes for 1×10^2, 1×10^3, and 2×10^4 electrons, respectively. **e** Normalized intensity profile of interference fringes in (d) integrated in *y*-direction. **f** Simulated intensity profile of interference fringes (equal to that shown in Fig. 2.19d) for comparison

where S_w is the area of the wire plane except for the wire region, showing this equation being equivalent to (5.26) for large n.

In Fig. 5.14, the interference fringes simulated with SLWs is shown. As shown in Fig. 5.14a, 15 electrons in the image plane were plotted along the SLW line traces and the red point on each line shows the electron position obtained by using random numbers for the estimation of the probability of locating electron positions. Simulated interference fringes are shown in Fig. 5.14b, c, and d for $n = 1 \times 10^2$, 1×10^3, and 2×10^4 electrons, respectively. Interference fringes appear with the increase in the number of incident electrons (see video in Fig. 5.15). A smooth intensity profile in (e) was obtained by integrating the intensities in (d) along the y-axis. It is noted thar this intensity profile is in good agreement with the interference fringe profile evaluated with the plane wave incidence (f) [11].

5.2 Spin

In the analysis of diffraction and interference phenomena with the SLW model described in Sect. 5.1, each incident electron is vaguely assumed to have a spinning motion. In this section, the spinning motion of the linear wave is discussed on the basis of the spin angular momentum and the spin magnetic moment in quantum mechanics.

Fig. 5.15 Simulated interference fringes. Video of simulation was prepared in collaboration with Mr. Tomita

5.2.1 Magnetic Moment of Electron Spin Interpreted with SLW Model

In Sect. 3.2.4, we presented reconstructed phase images showing the phase shift due to the magnetic field caused by the spin polarization of secondary electrons around the charged insulating materials [12]. To carry out the detailed simulation described in Sect. 3.2.4, we used quantized values characterizing the electron spin, i.e., the quantized spin angular momentum and spin magnetic moment, the magnitudes of which are $\hbar/2$ and μ_B ($= e\hbar/2m$: Bohr magneton), respectively. The precession of the spin magnetic moment under an external magnetic field corresponding to the two states $\pm\mu_B$ is illustrated in Fig. 5.16.

Figure 5.17a and b simply illustrate the linear wave in red and its rotation, respectively. Taking into account the quantized angular momentum along the spin axis for which the magnitude is $\frac{\sqrt{3}\hbar}{2}$ according to quantum mechanics, the spinning motion of the linear wave is illustrated in Fig. 5.17c, where the angle between the spin axis and the direction of the external magnetic field is 54.7° and the magnitude of the vertical component of the quantized spin angular momentum is $\hbar/2$. The arrows indicate the direction corresponding to the rotation of a negative charge clockwise from above.

Note that the linear wave rotates quickly around the central point. The assumption of de Broglie's relation for the mass (m) and frequency (ν) in (1.58) is satisfied at the central point only, while in other parts of the linear wave, the condition of (1.58) is not satisfied, resulting in a small contribution of the mass to the spin angular momentum and energy. Although the linear wave is long and its velocity increases with the distance from the central point, the angular momentum produced by the outer part of the linear wave is considered to decrease resulting in a quantized constant value for the angular momentum for which the vertical component is $\pm\hbar/2$.

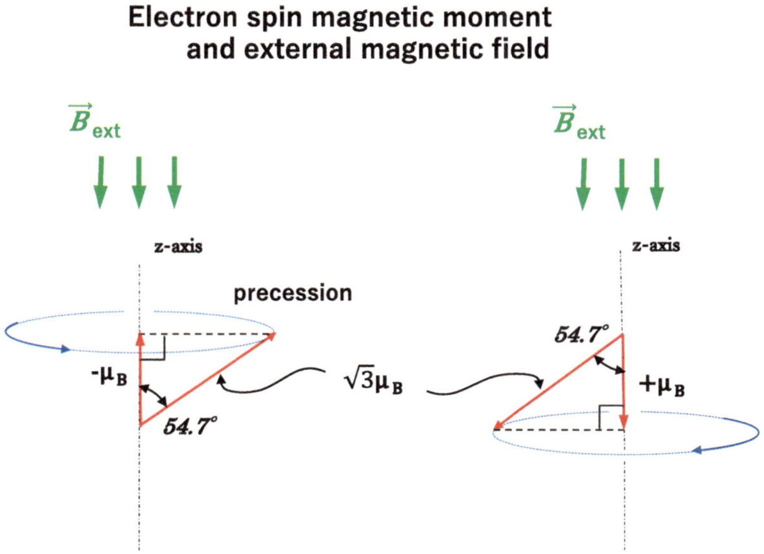

Fig. 5.16 Illustration of electron spin magnetic moment under external magnetic field

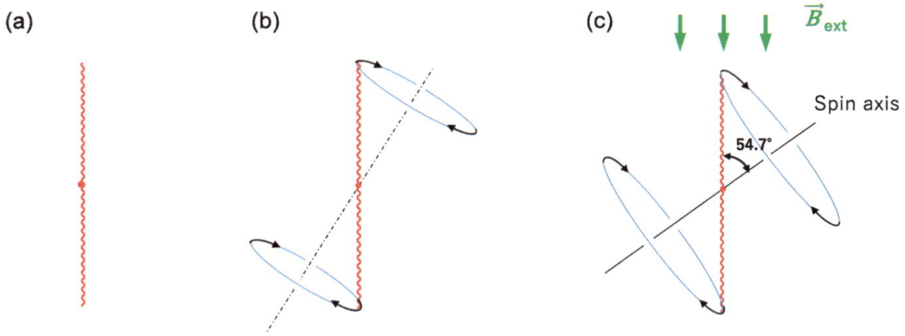

Fig. 5.17 **a** Formation of linear wave. **b** Rotation of linear wave. **c** Spinning motion of linear wave satisfying quantized spin angular momentum

Figure 5.18 shows the spinning motion of a linear wave. The direction of motion corresponds to the motion of a negative charge, as indicated by the arrows, and is the same as that shown in Fig. 5.17. The spinning motion of a linear wave clearly forms a cone.

Figure 5.19 illustrates precession of the spin magnetic moment under an external magnetic field. Two cones formed by the spinning motions are indicated by red arrows for positive current, and the directions of the presession are indicated by black arrows. (These illustrations of SLW motion are used in the animation of the change from the linear wave feature to the particle feature presented in Fig. 5.23). The magnitude of

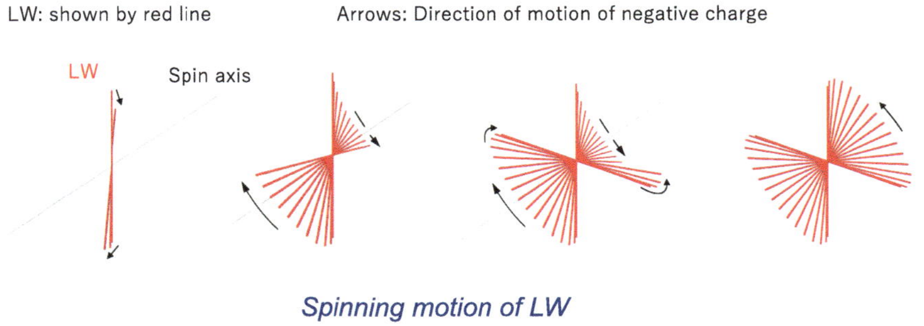

Fig. 5.18 Series of diagrams illustrating spinning motion of linear wave

the quantized magnetic moment along the spin axis is $\sqrt{3}\mu_B$. Due to the precession under the external magnetic field, the expected quantum mechanical values of the spin magnetic moment are $+\mu_B$ and $-\mu_B$ along the external magnetic field, as shown in Fig. 5.19, where the former (parallel to the direction of the external magnetic field) is more energetically stable than the latter (antiparallel to the direction).

5.2.2 Comparison of SLW Model with Top Spin Motion

To understand the basic features of electron spin and its precession under an external magnetic field, we compare the spinning and precession of a top to those of the linear wave, as shown in Fig. 5.20a and b, respectively.

In Fig. 5.20a, the top spins and precesses counterclockwise as seen from above. The angle between the spinning axis and the precession axis is θ_t, where θ_t is a variable. The vector from the fulcrum to the center of gravity is $\overrightarrow{r}$. The force of gravity $\overrightarrow{F}$ downward and the vector $\overrightarrow{r}$ form the torque $\overrightarrow{N_t}$; i.e.,

$$\begin{aligned}\overrightarrow{N_t} &= \overrightarrow{r} \times \overrightarrow{F} \\ &= \overrightarrow{r} \times M\overrightarrow{g},\end{aligned} \tag{5.30}$$

where M is the mass of the top and g is the gravitational acceleration. The direction of $\overrightarrow{N_t}$ corresponds to the direction of the precession.

In Fig. 5.20b, the red conical surface corresponds to the trace of the spinning motion of the linear wave around the spin axis. We assume that the spinning motion of the positive charge in relation to the direction of the spin magnetic moment is counterclockwise from above, which is opposite to that shown in Fig. 5.17c. Under these conditions, in comparison with the case for the top, the torque $\overrightarrow{N_s}$ is given as

$$\overrightarrow{N_s} = \sqrt{3}\mu_B \times \overrightarrow{B}_{ext}. \tag{5.31}$$

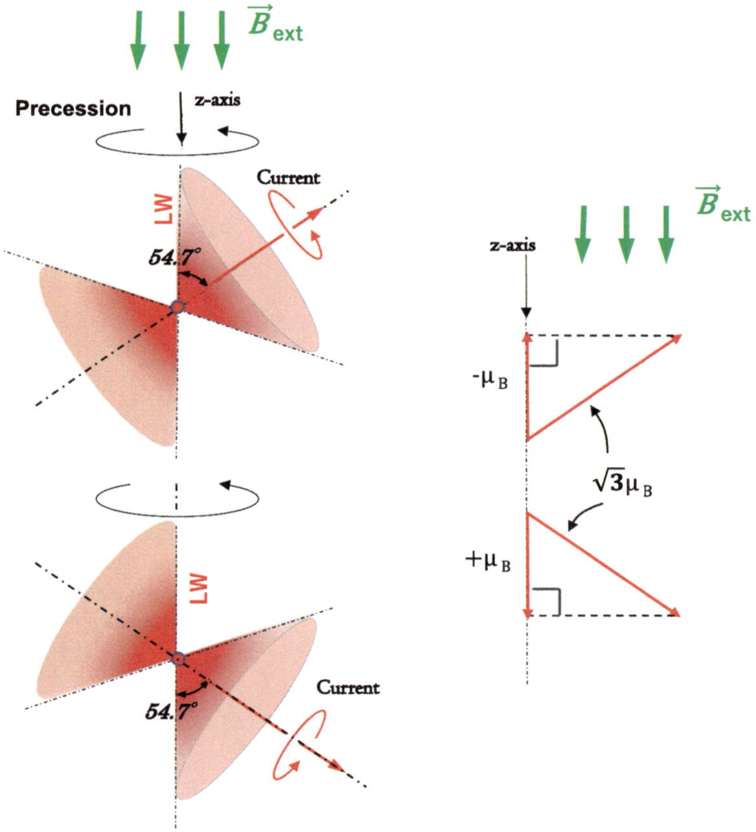

Fig. 5.19 Illustration showing precession of spin magnetic moment under external magnetic field

Thus, the precession of the electron spin is basically interpreted in a manner similar to that for the motion of the top. The unique quantized state of the precession with the fixed angle (54.7°) for the electron spin can be understood in terms of the wave properties, i.e., the maintenance of the waveform in the SLW model, as noted in the next section.

5.2.3 Interpretation of Larmor Precession of Spin Magnetic Moment with SLW Model

Now we discuss the characteristic feature of the precession of the electron spin, focusing on the importance of the maintenance of the waveform of the linear wave. As shown in Fig. 5.21, we selected four positions of the linear wave (indicated

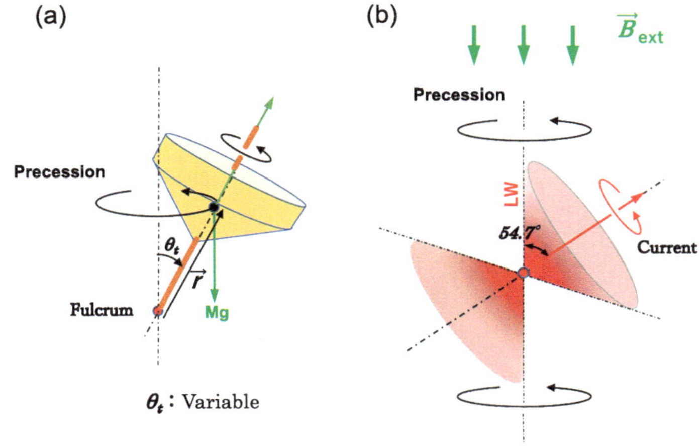

Fig. 5.20 Comparison of precession of top with that of SLW. Illustration prepared in collaboration with Mr. Tomita

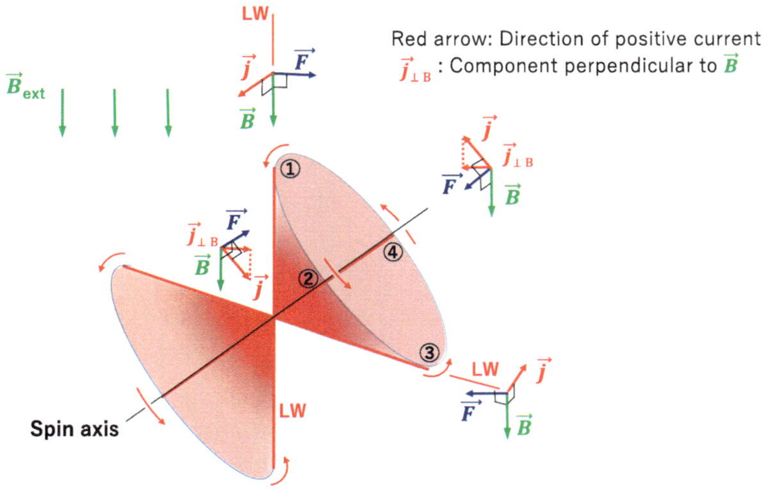

Fig. 5.21 Spinning motion of linear wave showing directions of electric current and Lorentz force under external magnetic field at four positions ①, ②, ③, and ④

by red lines) during the spinning motion, i.e., positions ①–④. For each position, three vectors for Fleming's left-hand rule, i.e., $\vec{F}$ (Lorentz force), $\vec{B}$ (magnetic flux density), and $\vec{j}$ (positive charge current) are shown. During precession, the linear wave always passes through position ① for each spinning motion, while at positions ②–④, the linear wave continually shifts in accordance with the precession. Thus,

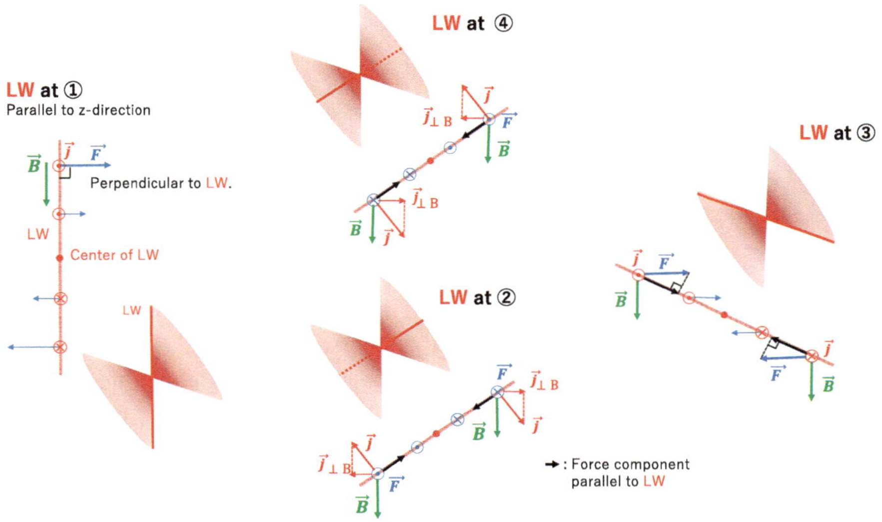

Fig. 5.22 Directions of electric current and Lorentz force under external magnetic field

the probability of the linear wave passing through position ① is the highest during precession.

The reason for this can be understood in terms of the maintenance of the waveform of the linear wave. The maintenance of the waveform facilitates understanding the diffraction and interference phenomena with the SLW model. Basically, a change in the direction of positive charge $\overrightarrow{j}$ produces a corresponding change in the direction of Lorentz force $\overrightarrow{F}$. At positions ① and ③, the forces are opposite, and similarly those at positions ② and ④ are opposite. In Fig. 5.22, the components of the force along the linear wave are indicated by black arrows. At position ①, there is no force along the linear wave while there is a large component of the force at ③ along the linear wave, which tends to disturb the maintenance of the linear waveform. Thus, the reason the linear wave tends to pass through position ① during precession with the force free is attributed to the maintenance of the waveform of the linear wave with a constant wavelength as the cases of the diffraction and interference phenomena caused by the maintenance of the waveform of the linear wave discussed in Sect. 5.1.5.

Figure 5.23 presents an animation (©RIKEN) based on the SLW model showing the linear wave feature changing into the particle feature through the precession of the magnetic moment of the electron spin. This animation was prepared in collaboration with Ms. Ishida and Ms. Terakura.

The characteristic points of the three-step processes (see Fig. 5.24) from the linear wave feature to the particle feature are summarized as follows.

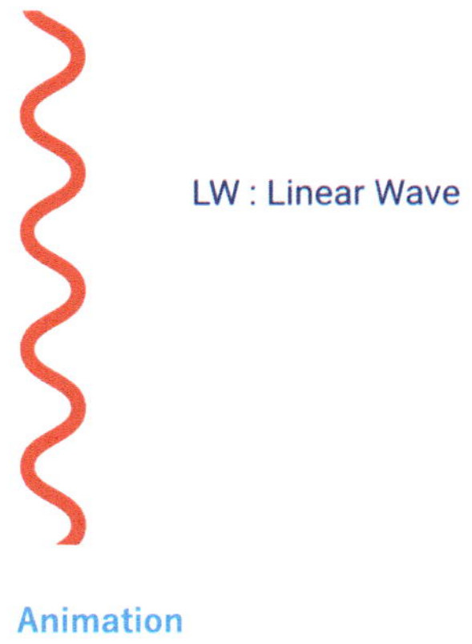

Fig. 5.23 Animation: Change from linear wave feature to particle feature. Animation (©RIKEN) was prepared in collaboration with Ms. Ishida and Ms. Terakura

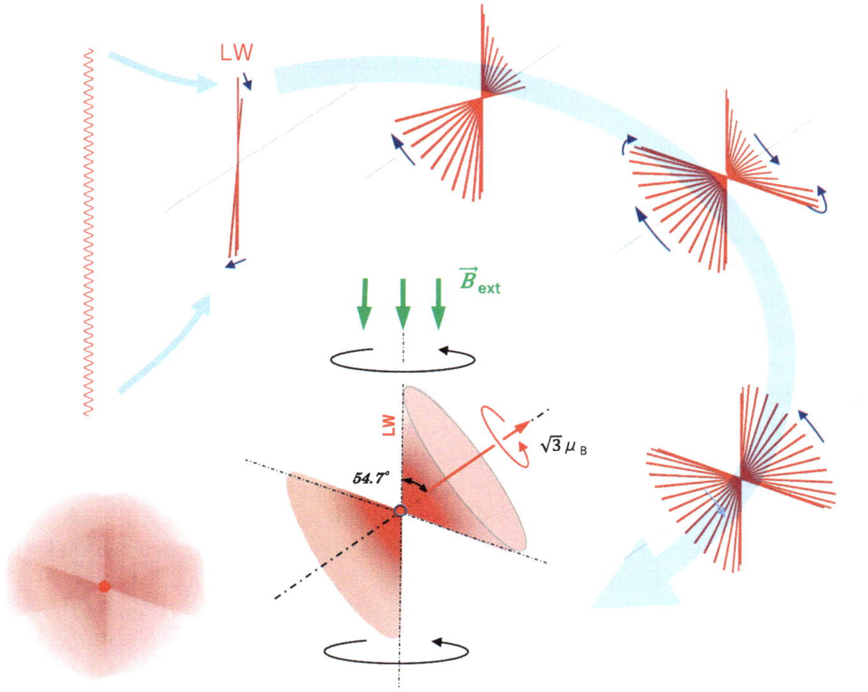

Fig. 5.24 Illustration showing processes corresponding to change from linear wave feature to particle feature

(1) Formation of linear wave

As derived by Einstein (1.57), the gravitational compressive force and electric repulsive force are considered to be balanced along the linear wave. According to the condition for the linear wave, i.e., (5.12), the deviation from Minkowskian metric tensor is small for large N_e. There is no singular point along the linear wave.

(2) Spinning motion of linear wave around spin axis

Avoiding the localization of the field at the specific region of the linear wave, it ends up spinning motion with quantized spin angular momentum $h/4\pi$. Due to the electric charge in the linear wave, the spinning motion results in the spin magnetic moment μ_B.

(3) Precession under external magnetic field

Quantized spinning motion of the linear wave forms the double conical surfaces (Fig. 5.24). These conical surfaces tend to rotate further showing the isotropic distribution of the field around the center of the linear wave. Under the external magnetic field, these conical surfaces make Larmor precession indicating the almost isotropically distributed field around the center of the linear wave as seen in the animation. In such motion, the waveform of the linear wave with a constant wavelength is maintained in a similar manner to other quantum phenomena such as diffraction and interference of the electron.

In relation to Riemannian space and Minkowskian space, the characteristic feature of the SLW is noted. In each process, there are no singular point. Through the three processes noted above, the time average of the space curvature becomes almost 0 and the field localization is suppressed resulting in the isotropic field formation around the central point of the SLW. In other words, Riemannian space (general relativity) around the SLW comes close to Minkowskian space (special relativity). Only at the central point, the regular oscillation of the field or curvature is maintained. The field oscillation at this point (red point in the lower left illustration in Fig. 5.24) is eventually considered to be connected to the rest mass energy as de Broglie postulated in (1.58), i.e., $mc^2 = h\nu$ on the basis of special theory of relativity.

We further note that the close correspondence between the quantum phenomena and the maintenance of the waveform of the linear wave. In the electron holographic observation of collective motion of electrons around the charged insulating specimens noted in Sect. 3.3.2, the existence of metallic elements was found to sensitively affect the distribution of the secondary electrons. In the SIM observation on the bulk mica surface in Figs. 3.60 and 3.61, the existence of metallic elements between the coated regions "e" and " - " separated at micrometer distance was found to significantly affect the movement of the electrons corresponding to the tunneling phenomenon. With the SLW model, this tunneling phenomenon is also explained as follows. In the insulating boundary region, there are irregular charge distributions. In order to keep the waveform with the constant wavelength, the motion of the SLW tends to be disturbed. On the other hand, the existence of the metallic elements tends to form smoother electric potential distribution at the boundary region than the case without the metallic elements. Due to the smooth electric potential distribution, the SLW easily keeps the waveform with the constant wavelength, and thus the disturbance of its motion tends to be suppressed resulting in the easy movement of the electrons between the coated regions corresponding to the tunneling phenomenon.

References

1. D. Shindo, T. Tomita, *Material Characterization using Electron Holography* (Wiley-VCH, Weinheim, 2023)
2. R.D. Heidenreich, *Fundamentals of Transmission Electron Microscopy* (Interscience, New York, 1964)
3. M. Born, E. Wolf, *Principles of Optics*, 2nd edn. (Pergamon Press, Oxford, 1964)
4. D. Shindo, Z. Akase, Mater. Sci. Eng. R **142**, 100564 (2020)
5. J.D. Jackson, *Classical Electrodynamics* (Wiley, New York, London, 1962)
6. A. Einstein, L. Infeld, B. Hoffmann, Gravitation equations and the problem of motion. Ann. Math. **39**, 65 (1938)
7. L. Infeld, A. Schild, Rev. Mod. Phys. **21**, 408 (1949)
8. J. M. Cowley, *Diffraction Physics*. Elsevier New York (1975)
9. D. Shindo, K. Hiraga, *High Resolution Electron Microscopy for Materials Science* (Springer, Tokyo, 1998)
10. J. Komrska, Scalar diffraction theory in electron optics. Adv. Electron. Electron Phys. **30**, 139 (1971)
11. D. Shindo, T. Tomita, *US 2023/0223232: Electron Microscope Analysis System*
12. T. Sato, K. Shimada. Z. Akase, H. Magara, T. Tomita, D. Shindo, *Microscopy* **73**, 367 (2024).

Index

D. Shindo and Z. Akase, *Electron Holography for Electromagnetic Materials*, Springer Series in Materials Science 347, https://doi.org/10.1007/978-981-97-3537-2

GPSR Compliance

The European Union's (EU) General Product Safety Regulation (GPSR) is a set of rules that requires consumer products to be safe and our obligations to ensure this.

If you have any concerns about our products, you can contact us on ProductSafety@springernature.com

In case Publisher is established outside the EU, the EU authorized representative is:

Springer Nature Customer Service Center GmbH
Europaplatz 3
69115 Heidelberg, Germany

Batch number: 10280114

Printed by Printforce, the Netherlands